the discovery of time

the discovery of time

TIME

edited by stuart mccready

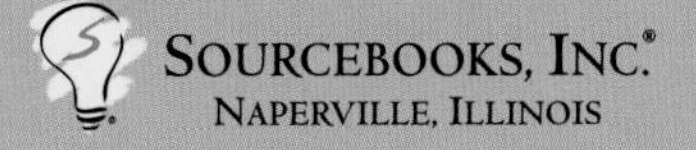
SOURCEBOOKS, INC.®
NAPERVILLE, ILLINOIS

contents

author biographies

Ralph Mistlberger

Writing on "Keeping Time With Nature" (Chapter Two), Ralph Mistlberger is in charge of the Circadian Rhythms Laboratory at Simon Fraser University, Burnaby, British Columbia. He investigates the brain mechanisms that control daily rhythms of waking and sleeping in mammals, and applies this knowledge to shift work and jet travel problems, as well as to questions about sleep. He has published extensively in professional journals.

Clive Ruggles

Professor of Archaeoastronomy in the School of Archaeological Studies at the University of Leicester, Clive Ruggles writes on "Prehistoric Timekeepers" (Chapter Three) and "Exotic Timekeepers" (Chapter Five). His research centers upon the astronomical significance of Neolithic and Bronze Age stone monuments of Britain and Ireland. He has also worked on indigenous calendars in many other parts of the world. His writing and editing work includes *Megalithic Astronomy*, *Records in Stone*, *Archaeoastronomy in the 1990s*, *Astronomies and Cultures*, and *Astronomy in Prehistoric Britain and Ireland.*

Robert Hannah

A graduate of Otago University in New Zealand, and of Oxford University, Robert Hannah writes on "The Moon, the Sun, and the Stars" (Chapter Four). He is Associate Professor in Classics at Otago, where he teaches Greek and Roman Art and Archaeology, and is Dean of the School of Language, Literature, and Performing Arts. He has published papers on the ancient Greek calendar, and on Greek and Roman astronomy.

Sara Schechner

Sara Schechner is the Curator of the Harvard Collection of Historical Scientific Instruments. She writes on "The Time of Day" (Chapter Six). Her interests range over material culture, science and religion, early scientific instruments, early modern history, popular culture and science, and the history of astronomy and physics. She was Chief Curator and Principal Gnomon Researcher at Silver Spring, Maryland, from 1996–2000, as well as serving as Adjunct Professor in the Department of History in the University of Maryland from 1995–98. Visiting Professor at Sarah Lawrence College, Bronxville, New York from 1991–92, she has been the Editor of Historic Scientific Instruments of the Adler Planetarium since 1984. Her highly acclaimed books include *Western Astrolabes*, and *Comets, Popular Culture, and the Birth of Modern Cosmology.*

Michael Roberts

Writing on "Journeys into Deep Time" (Chapter Nine), Michael Roberts studied geology at Oxford, and spent three years in Africa as an exploration geologist. After studying theology at Durham, he was ordained into the Anglican Church in 1974, and spent thirteen years in parishes in Liverpool before taking up his present position as Vicar of Chirk, near Llangollen, Wales, in 1987. He has written articles on science and religion (one on Darwin and Design received a Templeton Award in 1997) and Darwin's geology. He is married to Andrea and they have two grown children.

John Wearden

Professor John Wearden writes on "More Clocks within Us" (Chapter Ten). He conducts experimental and theoretical studies of timing in humans and animals at the University of Manchester, in the U.K. He teaches, among other subjects, the psychology of time and the psychology of space, time, and music. His research projects have included work on temporal processing in patients with Parkinson's disease, arousal and subjective time, and timing processes in children. He collaborates actively with colleagues in France, Germany, and Belgium.

Robin Le Poidevin

Robin Le Poidevin writes on "Puzzles about Time" (Chapter Eleven). He was born in Derby, England in 1962. He was an Exhibitioner of Oriel College, Oxford, where he read philosophy, psychology, and physiology. He was a research student at Emmanuel College, Cambridge, where he received a Ph.D. in 1989. From 1988–89 he was Gifford Research Fellow in Philosophy and Natural Theology at the University of St. Andrews, Scotland. Since 1989 he has been teaching at the University of Leeds, where he is now Senior Lecturer and Head of the School of Philosophy. His publications include *Change, Cause, and Contradiction* (1991), *Arguing for Atheism* (1996), *The Philosophy of Time* (1993, coedited with Murray MacBeath), and *Questions of Time and Tense* (ed.,1998).

Stuart McCready

The editor of this series is Stuart McCready. He also writes "Thinking about When It Is" (Chapter One), "Beyond Dead Reckoning" (Chapter Seven), and "The Triumph of the Clockmakers" (Chapter Eight). Stuart McCready has been involved with history across its entire span, from ancient Egypt to the twentieth century. Trained as a philosopher, he has taught in universities in Canada and Nigeria. He has also taught French and German in state and public boarding schools in England. During a thirty-year career in publishing, he has been executive editor on a wide range of books, covering psychology, human behavior, natural history, health and medicine, warfare, theater, astronomy, and art.

1: thinking about when it is

Stuart McCready

Getting the measure of time

The power of appearance leads us astray and throws us into confusion...whereas the art of measurement...would have caused the soul to live in peace and quiet abiding in the truth Plato

Gold enamel clock face, design begun by Mauro Coducci (1440–1504)

It is a piece of ancient Greek wisdom that counting and measuring things is a much surer path to knowledge and understanding than any other. The human discovery of time has not in any way contradicted this lesson. After many thousands of years we are no closer than we ever were to being able to say what time is. It has no shape, no smell. It leaves no mark of its own as it passes. It has no appearance by which we can know it. Yet we are now much better able to say what time it is. This book is the story of how the human race has learned, with ever more astonishing precision, to record, compare, and think about measured units of time—an aspect of our experience that seems otherwise indefinable. It is also the story of how our mastery of time has mastered us, by making it possible to demand of ourselves feats of synchronized living that our more distant ancestors could not have imagined.

Bringing time to mind

Measuring time is in our nature. As Ralph Mistlberger explains in Chapter Two, our bodies come equipped with an intricate biological mechanism for synchronizing our body rhythms with the daily cycle of light and dark that comes with the rising and setting of the sun. This "circadian clock" was already long in place, doubtless even in very early lifeforms, before prehistoric humans appeared on the earth to make time an object of consciousness. The most basic unit of time for these humans must surely have been the same day that was so fundamental to their biology.

Prehistoric cultures, however, have left little to tell us how they thought of this time. Was a day the period of light between sunrise and sunset, followed by another time unit—the period of dark between sunset and sunrise? Or did each daytime and nighttime together make a day? When did it start? Surely not at midnight—as our day does—for what could tell people in prehistoric times when they had reached the middle point of the dark time? Did it start at dawn, when light first breaks on the eastern horizon or, as in ancient Egypt, at sunrise (when the orb of the sun begins to appear)? It may have started at sunset, as is still the case in the Jewish and Islamic religious calendars. Noon was the beginning of the day in astronomical calculations from the time of Ptolemy in second-century Alexandria until 1925, when a midnight start was adopted by international agreement. The definition of any unit of time, even one so biologically inbuilt as the day, is subject to human convention, and we have no record of what conventions prehistoric peoples experimented with.

Some prehistoric peoples must have counted lunar months—the cycle from the new moon to full moon and back again. After the day, this would have been the easiest heavenly regularity to define. Since the lunar cycle takes about twenty-eight and one-half days, sometimes it would be twenty-nine days, sometimes thirty, before the evening sky told the village priest that a new month was starting. It is not surprising that when Egyptian and Mesopotamian civilizations appeared in history, about five thousand years ago, they already had working lunar calendars.

Prehistoric people must also have thought in years. Counting up months is not very good for calculating these, since the cycle of the seasons takes about twelve and one-half lunar cycles. The lunar cycle doesn't send any distinctive signal every year on about the same day, and so it offers nothing to mark a start of a year. Seasonal points on the horizon for sunrise and sunset do, however. Several prehistoric monuments, such as Stonehenge in England and Newgrange in Ireland, show apparent alignments to the most southerly point on the horizon where the sun can be seen rising in winter, or to the most northerly point of sunrise in summer. It is natural to guess that these alignments mean that such seasonal turning points were turning points in a calendar.

e Callanish standing nes are on Harris, ıter Hebrides, Scotland.

Yet no individual prehistoric artifact, such as marks on a piece of bone, or a standing stone seemingly aligned to a moonrise or a sunrise point on the horizon, is by itself a reliable relic of prehistoric timekeeping. Instead, there is always the very real possibility that its resemblance to a timekeeping device has arisen by mere chance. As Clive Ruggles goes on to explain in Chapter Three, this possibility can be eliminated only by patiently comparing the objects that a whole culture has left us. Is there a consistent pattern of making marks in patterns that add up to lunar cycles, or in the pattern of aligning tombs or stone circles with the summer or the winter sunrise?

Such patterns have been less clearly established than a more general one of orientation to the sun. What has emerged most clearly are patterns, in some cultures, of aligning tombs with points on the horizon where the sun rises, without any special favor to the summer or winter sunrise. One Stone Age Mediterranean culture aligned its tombs to any point on the horizon above which the sun can be seen climbing in the morning. Through their tomb alignments, a few Stone Age communities betray a special interest instead in the falling sun or the setting sun. It is with times of day, rather than times of the month or the year, that Neolithic architecture gives us its most concrete evidence of prehistoric synchrony. Just how these times of day mattered remains a mystery.

Fitting times together

Days, lunar months, and seasonal years are natural units of time that impress themselves on us through cycles of nature. There is an age-old human expectation that these three units should fit together neatly into a system, in which months are multiples of days, and years are multiples of months. In Chapter Four, Robert Hannah explains why this is not possible, and how successive adjustments in our expectations brought us to the most widely used calendar in the world today—the Gregorian.

Lunar calendars begin each month at a specific point in the lunar cycle (such as the appearance of the first crescent moon). To keep faith with this system of timekeeping, you have to give up the notion of a consistent seasonal year. Either the "year," made up of twelve lunar months, has a consistent length but falls short of a seasonal year, and so wanders backwards through the seasons (as in the Islamic calendar), or it is kept in line with the seasons by adding a thirteenth month from time to time. Solar calendars, such as the Gregorian, instead sacrifice the integrity of the month to that of the year. Because the months of the Gregorian calendar do not align with phases of the moon, and its arbitrary months are what the word "month" has now come to mean, we have to insert the word "lunar" in front to speak of months in their original sense.

The integrity of our day, the time unit nature picks out for us more clearly than any other, matters more to us than the integrity of our year. We would never accept a system that added extra minutes or hours to various days to make 365 days add up to the time it takes the earth to orbit the sun. This would result in days that wandered free of the rising and setting of the sun—a day would come when the sun rose at midnight on the equator. So the Gregorian calendar has irregular leap years, to add a whole extra day every four years, and it drops one day in three centuries out of four to make up for the fact that there is not quite an extra quarter day in the seasonal year. For the ancient Egyptians, by contrast, the integrity of their 365-day administrative year was as important to them as the integrity of the day. Consequently, they sacrificed seasonality, allowing their administrative calendar to drift. Schemes to introduce leap years were resisted up until Roman times.

The Gregorian calendar is what tells us when the children start school, when we should celebrate a birthday, and when we have to meet a colleague for an appointment. Because we all accept the way it organizes the days and the years, we can talk to each other without confusion about the date of a great event from centuries ago. The system works so well, and most of us have so little contact with alternative calendars, that Gregory's can seem like some absolute insight into the way God organized time in the beginning. Yet this calendar is just the best attempt to date (and still a necessarily imperfect one) to reconcile conflicting time signals from the changing seasons, the sun, the moon, and the stars. In Chapter Five, Clive Ruggles reminds us that alternative calendars have worked well for people outside the historical path that led to our own calendar, and also for people living today beyond the reach of its present-day sphere of influence.

Inventing the week

Cycles in the natural world led us to the time units we call a day, a month, and a year. Other time units were invented purely by humans: weeks, for example. Naming them after gods, the way the Romans did or the traditional languages of West Africa do, helps to make the days of the week seem like part of the furniture of the universe. But the number of days in a week is entirely a matter of choice, and the whole idea of having weeks probably reflects nothing deeper than the fact that it's handy to assign a handful of names to days in a recurring pattern. It makes it possible for people to say, "I don't know what the date is on Tuesday, but I'll see you then."

Long before the calendar reforms of the French Revolution, which tried to introduce a ten-day week, ancient Egyptian scribes had found this an administratively sound number to work with. The Romans settled on a seven-day week, with each day named for the god who governed its first hour, and they passed the system on, eventually almost to everyone. But others still survive alongside it.

A rural market in West Africa, for instance, will meet every four days on its particular market day (traditionally under the protection of the god who gives his name to the day). People keep track of two overlapping weeks. Knowing what day it is in the seven-day week tells you whether this day is for going to church, going to work, or going to the city for Saturday shopping. Knowing what day it is in the traditional week (for example, if you are in an Igbo-speaking region, knowing whether it is *Eke*, *Oye*, *Afo*, or *Nwko*) tells you which nearby village you should go to if you're looking for a country market. Often, the last syllables in village place-names will help you as well, as in *Enugwueke*, *Ezuoye*, *Abolafo*, or *Akpankwo*.

Dividing the day

Similarly, there is nothing written in nature that says we have to divide the day into twenty-four hours of sixty minutes with sixty seconds each. The Egyptian convention of dividing the night, and by analogy the day, into twelve hours was a matter of cultural circumstance. It reflected the fact that any given night can be so divided by the risings of some twelve or other of thirty-six chosen star-groups. It was the morning risings of these groups which the Egyptians associated with the thirty-six weeks of their administrative year. By contrast, our division of each twenty-fourth of the day into sixty minutes is a relic of a completely different system: Mesopotamian base-sixty arithmetic.

For more than three thousand years, the most commonly measured hour was not a twenty-fourth of a day. It was the hour you could keep track of by dividing the daylight into twelve equal episodes. Since the length of daylight changes with the seasons, so does the length of an hour. Sara Schechner's Chapter Six is devoted to sundials, the remarkable instruments that measured these natural hours. Into the eighteenth century, they were commonly used timepieces, and they were the standard for setting clocks, which were then unreliable. Sara Schechner shows how the sundials that survive can tell us what time meant to the people who made and used them.

Equal hours have advantages over flexible ones. For those paid by the hour, it's no good if an hour means something different in June than in January. Chapter Seven will explain how crucial it is to navigation that a standard hour and the time it takes the earth to turn through fifteen degrees (a twenty-fourth of its circumference) are one and the same. It means that a Portuguese sailor can fix his position in terms of degrees of longitude west of Lisbon if he can compare Lisbon time with his local time somewhere in the Atlantic. But this was of little help before accurate clocks.

A French shepherd finding south at midnight, from observation of the stars.

A concerted feat of astronomy eventually provided star charts and manuals that allowed navigators to read in the night sky what time it is on the Greenwich meridian. By then, however, clocks had overcome their shortcomings. Chapter Eight tells the story of this technological achievement, how it underpinned other technological successes, and how the clock came to symbolize time consciousness and time pressure.

Times beyond our reach

The final chapters of this book deal with aspects of time that we still find hard to digest, beginning with its unimaginably long extent. Descartes, in the sixth of his *Meditations*, points out how limited our imagination is as a tool of understanding. He contrasted picturing something in your mind (as a way, for example, of understanding the idea of a three-sided figure) with the more intellectual understanding that is our only recourse in dealing with a figure of a thousand sides: "When I imagine a triangle…I see those three lines with my mind's eye. However, if I wish to think about a chiliagon…I cannot imagine a thousand sides…" As Descartes wrote this, the scientific revolution of which he was a part was turning its attention to the age of the earth. His mental experiment can be applied with equal effect to illustrate the challenge this would bring.

The new times that the clock brought to consciousness are easy to imagine: count one and that's a second; count sixty at the right pace and that's a minute. We also have some image or personal feeling of what a

The long backward exte[nt] of time was partly revealed through fossils, such as this shrimp from the Jurassic age (208 to 146 million years ago).

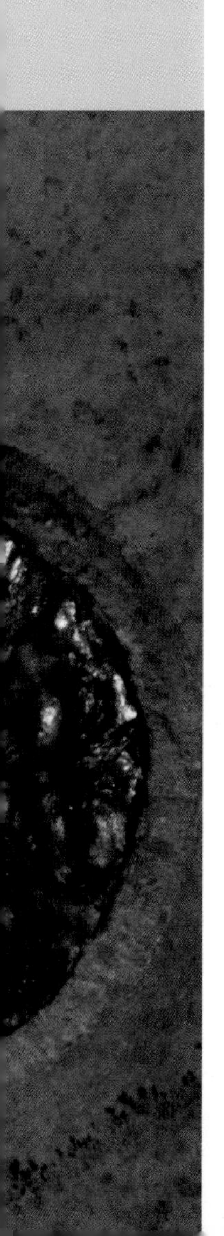

week, a year, or a lifetime is like. In turn, you can have many lifetimes in story form—perhaps six or seven thousand years of them, pictured in reliefs above the lintels of churches, or on mosaics and frescoes within. By contrast, as Michael Roberts recounts in Chapter Nine, the vast extent of time that Western civilization would now be asked to take into consciousness produced a vertiginous effect.

Other civilizations had calculated staggeringly long histories for the world. The Sumerians believed that before the Flood, eight kings had ruled over them during a fabulous time that lasted 241,200 years. The Babylonians, the Greeks, and the Indians shared a concept of the Great Year, a complete cycle of the cosmos, calculated as lasting millions of years. Much of the Christian world in the seventeenth, eighteenth, and nineteenth centuries based its count of the years since Creation on information from scriptures. This made the earth less than six thousand years old. Coming to terms with the new chronologies was a wrenching experience, and one that we have never fully absorbed. A useful therapy is to walk with Michael Roberts from the bottom of the Grand Canyon to the top, meditating on the significance of each rock stratum you pass through.

Time often becomes imponderable even in small units. What happens to time when we lose track of it for several minutes? Why does it move so slowly when we want it to go quickly? In Chapter Ten, John Wearden reveals what psychology has learned about the perception of time—how in addition to our circadian clock, further internal clocks (which may at times run disconcertingly slow or fast) help us to gauge where we are in time.

Finally Robin Le Poidevin confronts us in Chapter Eleven with puzzles about time. Beginning with paradoxes by which Zeno of Elia has tested the human mind for twenty five hundred years, he lays out questions that perplexed, among other great minds, those of Aristotle, St. Augustine, and Immanuel Kant. The effect of these puzzles may be to leave you asking how time is possible, or at least, how time is possible without the imposed conventions of a species bent on measuring things.

HORTUS CLIFFORTIANUS
TURNERA e petiolo florens foliis
serratis. Hort.Cliff.

2: keeping time with nature

Ralph Mistlberger

Our twenty-four-hour biological clock

A clock made with wheels and weights observes all the laws of nature...Likewise, I think of a human body as some kind of machine made from bones, nerves, muscles, veins, blood, and skin René Descartes

On May 21, 1989, twenty-seven-year-old interior decorator Stefania Follini, of Ancona, Italy, climbed out of Lost Cave, New Mexico. Stefania had spent 130 days there, completely alone. She had also broken the world record for isolation underground in an environment without any of the twenty-four-hour time cues by which we normally organize our lives. Stefania was a willing volunteer in an experiment designed to simulate long-distance space travel, to understand more about the effects of prolonged isolation on the human mind and body.

The human clock

Without knowledge of the time of day, the Italian woman slept and woke according to her own natural rhythms. In normal life, the duration of her sleep-wake cycle was twenty-four hours, matching the day-night cycle. Without time cues, her cycle lengthened immediately to twenty-five hours. Within a few weeks, it lengthened further, to as much as thirty-six hours. But not only had her daily rhythms changed dramatically, so too had her perception of time. When the experiment ended after 130 days, she thought that only about eighty days had passed.

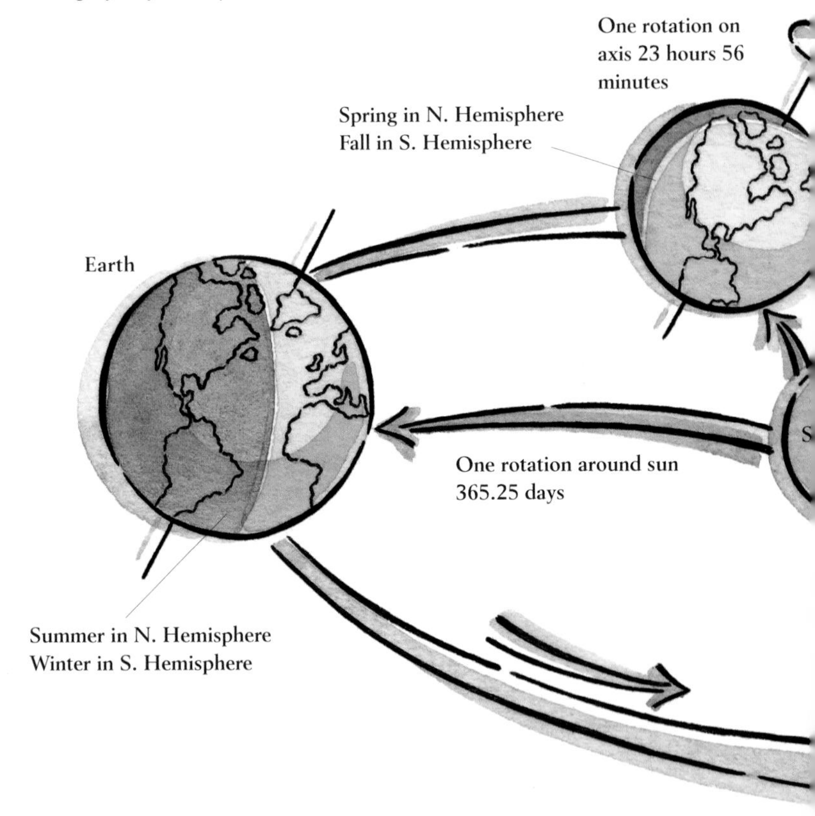

Over the past four decades, hundreds of people have been observed under similar conditions. These experiments, along with studies of plants, a wide variety of animals, and even bacteria, reveal that most organisms are innately rhythmic.

A question of good timing

Rhythms are a pervasive feature of life on earth. As the earth circles around the sun, it is also rotating all the time on its own tilting axis. This pattern of movement creates a twenty-four-hour solar day marked by dramatic daily cycles of light, temperature, and humidity. It also generates seasonal changes in climate and weather.

These daily and seasonal cycles present a formidable challenge to living organisms. Not surprisingly, most known species exhibit daily rhythms in their behavior, physiology, and biochemistry that are synchronized with the solar day. Many organisms also display seasonal rhythms, and some even exhibit rhythms related to the tides or to the cycles of the moon.

In humans, rhythmic patterns are most obvious in our daily sleep-wake cycle. People the world over prefer to sleep at night and be active in the day (with time set aside for an afternoon siesta in some countries). This daily cycle is deeply embedded in human biology. Metabolic rate, core body temperature, heart rate, blood pressure, digestion and excretion, hormone synthesis and release, immune system factors...all exhibit a twenty-four-hour rhythm. How are these biorhythms controlled, and do they help or hinder us as human beings?

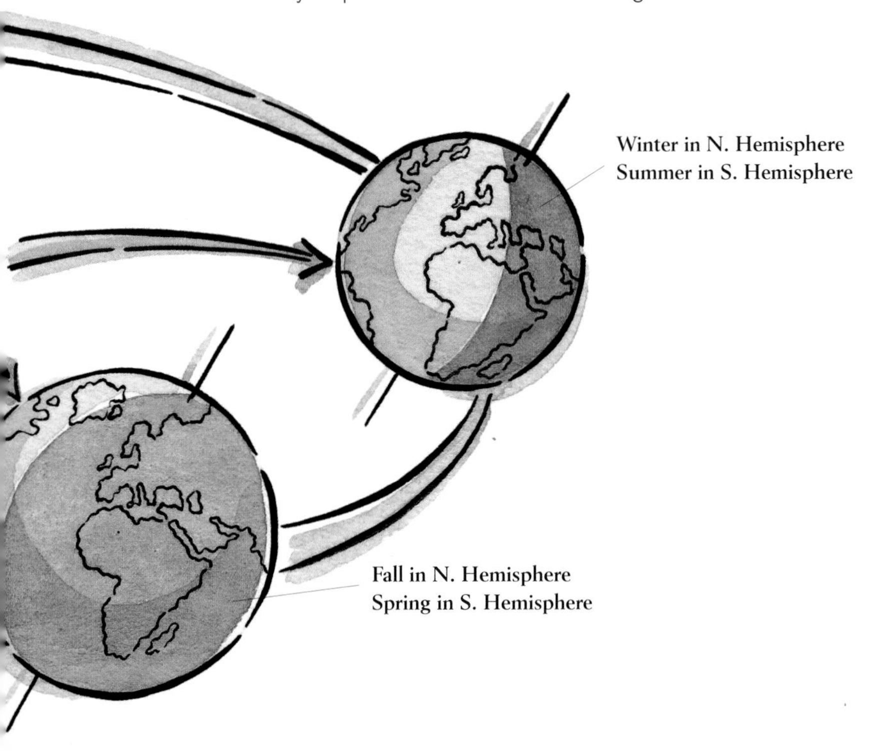

To answer these questions, we should think first about how biological rhythms and biological clocks relate to each other. A rhythm can be defined as any process that repeats itself at approximately regular intervals. But just because a living thing in its natural habitat displays a daily rhythm, that does not automatically mean that it possesses an

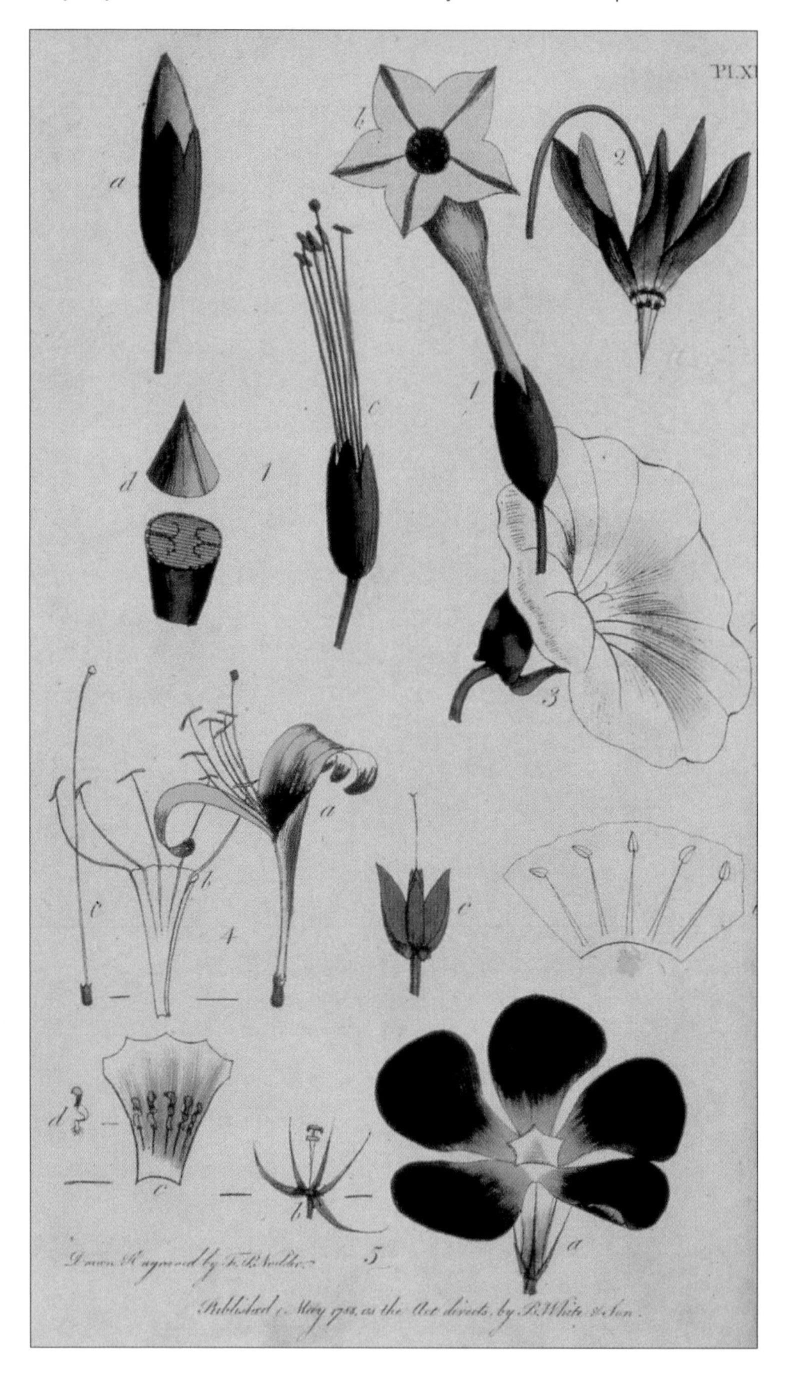

Eighteenth-century rendering of plants. The study of botany would sh light on the body clocks human beings.

internal biological clock—it may simply be responding directly to daily changes in light, temperature, or other external stimuli. To establish that a rhythm is truly inbuilt, it must be shown to persist when all the normal external time cues are taken away.

You may be surprised to learn that our first insights into internal rhythms were gained by studying the movements of plants. Groundbreaking experiments carried out on mimosa plants by de Mairen in the 1700s showed that their daily cycles persisted in the dark. However, at the time no one seems to have suggested that this was the result of an internal clock. Other environmental factors—such as subtle daily changes in temperature, humidity, or some unknown "X factor"—could just as easily provide timing information that plants might sense. In fact, de Mairen's discovery was slow to be accepted, and for many years the persisting daily rhythms of plants in the dark were attributed to unknown environmental stimuli that he and later experimenters must have failed to eliminate.

Biorhythms in the plant world

The first written records to mention daily cycles displayed by plant leaves and flowers date back to Alexander the Great (400 B.C.) and Androsthenes of ancient Greece. However, there is no record of anyone attempting to understand these cycles for many centuries to come. This is probably because the answer seemed self-evident: plants raise (or lower) their leaves and open their flowers during the day because they are responding directly to sunlight. Written evidence that an internal twenty-four-hour clock may be involved did not appear until the 1700s. This was also the century when Swedish botanist Carolus Linnaeus used plant cycles to create a twenty-four-hour "flower clock," having noted how flowers open and close at certain times of day.

In 1729, the Royal Academy of Sciences in Paris heard the results of a simple yet groundbreaking experiment that had been conducted by, of all people, an astronomer: Jean Jacques d'Ortous de Mairen. De Mairen was curious about the source of the daily leaf movements of mimosa plants (the leaves of which are notoriously sensitive), and so placed some in the dark for several days, noting the position of their leaves at regular intervals around the clock. He found that the daily movement rhythms persisted, as if the plants could "feel the sun without seeing it in any way."

De Mairen's work is the first known demonstration of a daily rhythm continuing in an unchanging environment. Since then, many species of plants and animals, including humans, have been shown to exhibit repeated daily rhythms in constant conditions.

Darwin and plant power

By 1850, the English naturalist Charles Darwin had become convinced that the daily rhythms of plants were inbuilt, but he appears to have misconstrued their significance. In his 1880 book, *The Power of Plant Movements*, he suggested that plant rhythms were due to "an effect on the seeds of long-term exposure to climate." Apparently he did not consider daily rhythms to have emerged by natural selection, as an adaptation that improves the survival and reproduction of plants in their natural habitats. Instead, he seems to have viewed them as a mere biological curiosity. This failure to consider daily rhythms as potentially useful may have contributed to the reluctance of the scientific community to take the concept of inbuilt rhythms seriously.

By 1930, the evidence for daily cycles had become difficult to explain without reference to an inbuilt biological clock. First, de Candolle (1778–1841), and later Pfeffer (1845–1920) and others showed that the persisting rhythms of plants kept in constant darkness are "circadian." This means that the cycles drift, rather than remaining absolutely constant, and the average length of one cycle only roughly matches the twenty-four hours it takes for the sun to reappear overhead each day. As a result, the rhythms of plants that are kept isolated from any time of day cues gradually lose their synchrony with the outside world.

It was also found that individual plants of the same species may have slightly different drifting cycle-lengths, and so become desynchronized from each other within just a few days. All this makes it very unlikely that the plants are actually responding to some unknown environmental factor, because it would suggest that not one, but many mystery factors are at work. Also, why would an individual plant be influenced by one factor but not the others?

As research into this field continued, scientists not only discovered that animals and plants harbor biological clocks, but also that their cycles contribute to reproductive success. Linnaeus showed that many flowers open their petals at very specific times of day. Bees forage for nectar in flowers, and the best time to forage is when a flower has just opened, before other insects and birds have had their fill. So, a bee's life would be much easier if it were able to recognize and remember the time of day that various flowers open.

Early in the 1900s, the Swiss neurologist Forel noted that honey bees indeed "seemed to have a good sense of time," based on the fact that hungry bees would arrive at his table just before the usual breakfast hour. Controlled laboratory studies by Stein-Beling (1935) and von Frisch (1950) confirmed Forel's findings; that bees could remember the exact time of day that they had received nectar, without any help from external time cues.

These studies also showed that bees could locate a food source by using the "azimuthal" position of the sun—the point on the horizon directly below the sun—as a compass. This is a remarkable ability, because the azimuth changes by fifteen degrees per hour as the sun steadily tracks across the sky. To use the azimuth as a directional cue, the bee must know the time of day, and make continuous adjustments in its direction of movement relative to the azimuth. Studies conducted by Kalmus in 1935 and Kramer in 1950 demonstrated that birds can also use the sun and stars to navigate in this way. This was the first really convincing evidence that animals have a twenty-four-hour internal clock.

Lindauer's research in 1960 charted bees at two feeding places. The second feeding place was six hundred and fifty-six feet (two hundred meters) south of the first. The bees regularly visited the two sites at specific feeding times, when food was available. The same results are obtained in controlled laboratories in which there are no time of day cues, demonstrating that bees have an inbuilt sense of time.

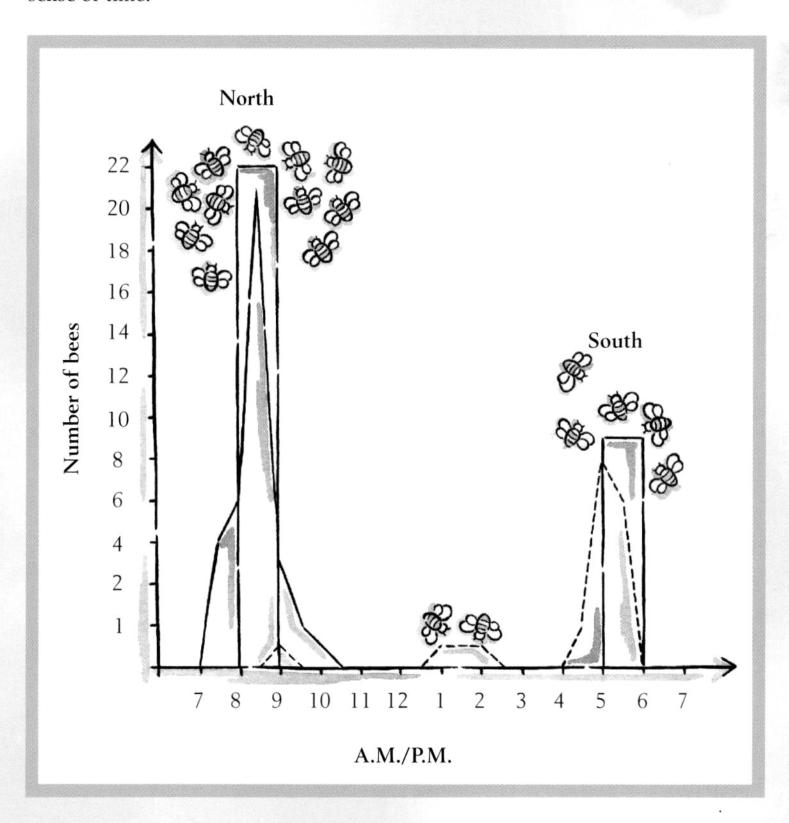

Circadian clocks

In 1954, Gustav Kramer spelled out a link between the "sun-compass" clock of birds and bees, and the persisting rhythms of plants and mammals that were isolated from any time of day cues. It looked like this same "circadian clock" might control daily rhythms and allow sun-compass orientation. Klaus Hoffman was soon to prove that this is so. He showed that birds in constant light have circadian rhythms with a period of about twenty-three and one-half hours, and demonstrated that, under these conditions, the birds make errors in sun-compass navigation that can be predicted exactly—based on their circadian cycles. The circadian clock is used not only to generate daily rhythms of behavior and physiology, it also provides precise time of day information necessary for sun-compass orientation. From these studies came widespread acceptance by the late 1950s of the idea that plants and animals have internal circadian clocks—also called "pacemakers"—that enable them to recognize time of day and coordinate their behavior with predictable daily changes in their environment. If a twenty-four-hour biological clock is to work effectively in this way, it must have two characteristics. First, it must keep reliable time. Second, it must be synchronized to environmental (local) time. Just imagine a wristwatch that kept changing speed and could not be reset to the correct time of day; it would quickly be of little use.

To achieve these goals of stability and reset-ability, the circadian clock has evolved two features: sensitivity to light and lack of dependence on temperature. A fundamental principle of biochemistry is that the rate of biochemical reactions depends on temperature; the higher the temperature, the faster the rate. If the rate of oscillation of the circadian clock were also temperature-dependent, then the clock would make a nice thermometer, but a poor timekeeper. Temperature independence is particularly vital for single-celled organisms, plants, and cold-blooded animals, whose tissue temperature is determined mostly by their surrounding environment.

We do not yet know exactly how this temperature independence works at a molecular level, but it is now well established that the time-period of circadian rhythms in most known organisms is only slightly, if at all, affected by their tissue temperature. The current thinking is that the clock's biochemical timing works in such a way that it compensates for temperature-induced changes in reaction rates. This means that the circadian cycle remains stable across a wide range of temperatures.

Although the time-periods of the circadian clock are not substantially affected by temperature, each period is only *approximately* twenty-four hours. The solar day is much more exactly twenty-four hours. How then does the clock avoid systematically drifting out of alignment with local time? It does so by virtue of its sensitivity to light. Around the time of dawn and dusk, and through most of the night, the clock can be reset by light.

Seeing the light

Imagine a human whose circadian clock runs naturally slow at twenty-five hours, rather than at exactly twenty-four hours. This person would naturally want to go to sleep and wake up each "day" one hour later than the previous day. That is, their personal, "subjective", day would in fact be twenty-five hours long. So, in order to catch up and remain in step with the twenty-four-hour world, this person's slow internal clock must be shifted forward by one hour each day. Conversely, if the person's internal clock were to run naturally fast, to a twenty-three-hour subjective day, then it would have to be reset backward by one hour each day.

So, how does the required resetting occur? This is where the circadian clock's sensitivity to light comes in. Scientists have found that the internal human clock, like the internal clock of many living things, is designed so that exposure to early morning light produces a shift forward in the circadian cycle, while light exposure around dusk produces a slight backward shift. During most of the day the clock remains largely insensitive to light. The net effect is a small daily adjustment of the internal clock to cope with the difference between its subjective daily cycle and the twenty-four-hour solar day.

In humans, the changeover point from the delaying shift to the early morning advancing shift takes place around the time when the core body temperature is at its lowest daily point. In most people, this low-point occurs two to four hours before their usual wake-up time. Understanding this property can help enormously when coping with things such as jet-lag or a dramatic change in daily routine.

Going backward and forward

Many people traveling across time zones or undertaking shift-work have problems resetting their internal clocks. They find themselves suddenly out of alignment with the external world. Here is what you should do. If you wish to shift your clock forward, to a later hour (for example, when traveling across time zones in an easterly direction), then seek out bright light early in the morning, but no earlier than about three hours before your usual wake-up time. If you wish to shift your clock backward (say, when traveling west), then you should seek out bright light during the first half of the night, particularly during the three to four hours leading up to your body temperature minimum. For your internal clock to adapt fully to a new time zone or work schedule, it may take as much as one full day per hour of shift.

All animals, not just humans, have a circadian clock—which works differently for those active during the day and those active at night.

A question of degree

We have seen that the timing of light is crucial for resetting the circadian clock, but the intensity and duration of light are also important. Nocturnal animals, to give one example, can be remarkably sensitive to light. The nocturnal Syrian hamster can keep to a twenty-four-hour light-dark cycle in which its exposure to light lasts for just one second per day! The internal clocks of animals that are active during the day, such as humans, are thought to be much less sensitive.

Early studies conducted by Jurgen Aschoff and Rutger Wever at the Max Planck Institute in Germany suggested that the human clock is especially insensitive to light, and is strongly influenced by social factors. Later work suggested sensitivity to very bright light of five thousand lux or more (comparable to the light at dawn), but more recent studies, conducted by Charles Czeisler, Richard Kronauer, and colleagues at Harvard Medical School, show that even typical room light in the one hundred–two hundred lux range can significantly shift human rhythms. However, the basic rule remains: the brighter the light, the larger the cycle shift.

At extremely high intensities, light may actually appear to stop the clock. This phenomenon has been seen in simple organisms such as the fruit fly *Drosophila melanogaster*. The Czeisler and Kronauer group has recently found that if an experiment is set up so that a period of six hours of very bright light overlaps with the body-temperature low-point for one or two circadian cycles, this can flatten human circadian rhythms just as if the clock had stopped. A third six-hour light exposure the next day then "restarts" the clock at a very different point within the cycle, as much as twelve hours adrift from the original cycle. No one can agree about the exact significance of this, but it is certain that the human clock can be reset substantially if light of sufficient intensity is applied at the appropriate point in the circadian cycle.

Not precisely twenty-four hours

Why the internal clock's cycle should be "circadian" rather than a perfect match for the twenty-four-hour cycle of light and dark is not known for certain. There are, however, some fascinating theories.

First, it makes sense that as long as there is a mechanism for resetting the clock, it is not essential that it should run to a precise twenty-four-hour cycle. In fact, no biological process can be expected to proceed for long without error, so even if the clock did have an exact twenty-four-hour timetable, one would expect it to run a little fast or slow from time to time. With its ability to reset itself, the clock can compensate for these errors and quickly get back to its normal state, in step with the solar day. So, in evolutionary terms, there was probably no selection pressure to evolve an infallible twenty-four-hour clock.

Some species may well have good reason to possess a clock that runs slightly, but systematically, faster or slower than the solar day. A fast-running clock should make an animal wake up early, and, as we all know, it is the early bird that gets the worm.

More generally, having a clock that runs systematically a little faster or slower than twenty-four hours may actually make synchronization between the clock and the day-night cycle more stable, on average. It might also help to preserve the timing of our daily wake-up relative either to sunrise (in animals active by day) or sunset (in nocturnal animals), which changes with the time of year. It is worth noting that the circadian clock in some animals appears to be almost exactly twenty-four hours, so there is no need for the clock to run a little slow or fast.

A final note about the "circadian" nature of the human clock. You may find in textbooks a statement that the human circadian clock runs at about twenty-five hours when isolated from any time of day cues. However, that estimate appears to be a product of the way human isolation studies have typically been conducted.

In most studies to date, people either controlled their own lighting, or were kept in constant light, as humans cannot be kept in *total* darkness for long. When the subject sleeps, the lights go out, one way (lamp off) or the other (eyelids shut). When the subject awakens, the eyes open and are exposed to light. As discussed above, light can shift the clock. It has recently been shown by the Czeisler and Kronauer group that people operating a self-controlled, light-dark cycle actually slowed down their clock. When this influence is taken away from the individuals, the natural human clock cycle turns out to be much closer to twenty-four hours—around 24.2 to 24.3 hours. This range is now believed to be the best estimate for the human circadian clock.

Of course, this is only a general rule, and some people's clocks run a little slower or faster than the average. These differences are thought to explain, at least in part, those familiar human types, the "early bird" (the early riser with a faster-running clock) and the "night owl" (the late riser with the slower-running clock).

When cycles break down

Stefania Follini, the record-breaking cavewoman who began our story, had a near twenty-four-hour sleep-wake cycle for only part of the time she spent underground, isolated from time of day cues. At other times, she exhibited very long cycles of up to ten hours continuous sleep followed by up to twenty-six hours of continuous waking.

Dramatic lengthening of the sleep-wake cycle was reported in some of the first human experiments of this type, dating from the early 1960s, and most if not all subjects who spend sufficient time in isolation exhibit this phenomenon. However, the underlying mechanism remains a puzzle because the sleep-wake cycle is the only major rhythm to be disrupted when people are isolated in this way. Most of the body's physiological rhythms—including its core body temperature—continue to follow an inbuilt cycle which is close to twenty-five hours. The traditional interpretation of this phenomenon, suggested by Aschoff and Wever, is that humans have not one but two circadian clocks: a "strong" clock that controls body temperature and has a very stable cycle of twenty-four hours, and a "weak" clock that controls sleep-wake patterns, and that has a more elastic cycle of up to fifty hours. Normally, the theory goes, the two clocks are coupled, with the "strong" pacemaker dominating, but in abnormal environments the coupling fails. Another interpretation is that these cycles are affected by instructions given to subjects prior to the experiments, or by the way in which the data collected was analyzed. In many of the studies, the subjects were asked to avoid napping. However, many subjects napped nonetheless and Aschoff and Wever, for example, excluded from their final data analysis any sleep that the subjects described as a nap.

More recently, Jurgen Zulley and Scott Campbell reanalyzed some of this data and found that most of these subjects actually did obtain some sleep whenever the body temperature rhythm was at its low-point (i.e., once in every roughly twenty-four-hour period).

Sometimes the subjects called this sleep a nap rather than their "main" sleep, even if the "nap" turned out to be six to eight hours long! When these so-called naps were included in the analysis, the apparent desynchronization between the sleep-wake and body temperature rhythms disappeared. Both rhythms cycled together within the same period of about twenty-four hours.

Other subjects remained awake at some temperature low-points. These people may have resisted sleep at certain low-points because they weren't sure that it was time to retire for the "day," and were trying to obey the instructions not to nap. In so doing, they may have ignored signals from their own internal clock that it was time to sleep. Some may simply have been engrossed in an activity, like reading a good book.

Losing track of time

Although this reanalysis is intriguing, other aspects of the subjects' behavior still need explaining. First, why did some of them call certain very long sleeps a "nap"? Second, during long sleep-wake cycles of thirty-six to fifty hours, the subjects usually ate only three times, calling one meal breakfast, one lunch, and one dinner. This means that, on a fifty-hour "subjective" day, meals were separated by up to twelve hours, although the subjects seemed unaware of this.

In general, the subjects systematically underestimated the passage of time by at least 50 percent. When asked to press a button at one-hour intervals, their estimates were usually around the ninety-minute mark. This would explain why Follini and most other subjects underestimated the number of "real" days that they spent in isolation.

The standard explanation has been that our sleep-wake pattern is controlled by a flexible internal clock that also controls the way in which we perceive long intervals of time (in the hours to days range). So, each time our sleep-wake pattern changes, we have a different perception of how time is passing.

However, this cause-and-effect may run the other way around. Something about being isolated from time of day cues might actually change our perception of time, and it could be this that causes fluctuations in our sleep-wake patterns. Subjects that mis-estimate the passage of time may voluntarily ignore the internal clock signals that normally cue the onset of sleep or waking. This would produce some very long sleeping and waking periods. The fact that this phenomenon is not seen in other species seems to suggest it is related to the human psychology of time perception, and the human capacity to willfully disregard internal sleep signals and stay up all night.

If it really is the perception of time that changes our sleep-wake cycle, then this has two important implications. First, it seems to make the existence of two circadian clocks—that is to say, one for temperature and another one for the sleep-wake cycle—unlikely. Instead, a single clock would explain all rhythms of about twenty-four hours, while human will and "faulty" time perception would cause changes in the sleep-wake cycle. Second, it would mean that this single circadian clock is not involved in time perception; it keeps a stable, inbuilt cycle of roughly twenty-four hours even when our perception of hourly and daily intervals fluctuates dramatically.

Clocks on the brain

Is there really a twenty-four-hour clock somewhere in our body? The behavioral evidence compiled over the past couple of centuries is compelling, but has it actually been identified, physically, in the brain?

In the 1960s, eminent behavioral biologist Curt Richter set out to discover the site of the clock.

Using rats, Richter systematically tested whether dissecting various organs and damaging parts of the brain affected their circadian rhythms. He found only one site at which damage appeared to disrupt the clock. This site was somewhere in the hypothalamus, a part of the brain linked with regulating body weight, fluid balance, core body temperature, sleep-wake patterns, and reproduction.

By 1972, two groups (led by Moore and Eichler at the University of Chicago, and Stephan and Zucker at Berkeley) had discovered the precise site of the clock. It is located in a small brain region known as the suprachiasmatic nucleus, or SCN, because it lies on top of the "optic chiasm," near the base of the brain at a point where the optic nerves cross as they enter the brain (see diagram on following page).

The SCN has the distinction of being the first place in the brain that receives information about light and dark from the retina of the eye. A direct link with the retina is something that we would expect from the clock, given that it is tied in with daily light-dark cycles. When cells in the SCN were selectively destroyed by applying small electrodes directly, the rats immediately and permanently lost their circadian rhythms. In the 1980s, researchers showed that rats whose SCNs had been destroyed regained their circadian rhythms if they received an SCN transplant from a donor rat. Similar results have been obtained with mice and hamsters. This is one of the best examples of recovering a lost brain function by transplantation of brain tissue.

Now we know where the clock is, how does it work? It may be that the clock relies on information being passed along a network of brain cells, as many cells as it takes to time a twenty-four-hour cycle. Alternatively, the clock may be contained entirely within single brain cells, although the cells would still have to "talk" to each other to make sure that they remain synchronized.

A side view through a mammalian brain, illustrating the position of the suprachiasmatic nucleus (SCN), the site of the circadian clock. Light entrains (synchronizes) this clock by activating photoreceptors in the retina at the back of the eye, which then send this information directly to the SCN via the optic nerve. The SCN resides just above the optic chiasm (OX). The SCN in turn distributes timing information to many brain systems that regulate our physiology and behavior.

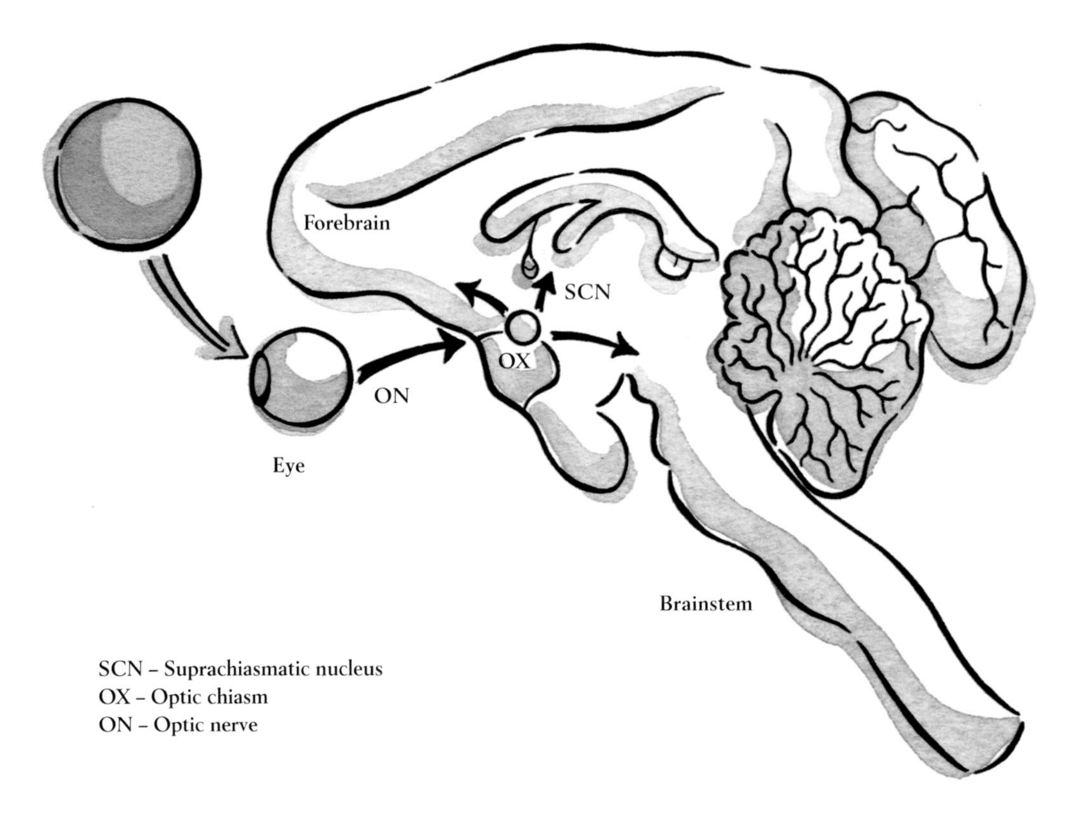

SCN – Suprachiasmatic nucleus
OX – Optic chiasm
ON – Optic nerve

Clocks within clocks . . .

It has been known for some time that single-celled organisms such as green algae have circadian rhythms, so their internal clock must be contained within single cells. In 1995, David Welsh, then a doctoral medical student at Harvard, proved that this is also the case with mammals. After five grueling years of research, Welsh managed to record the electrical activity of SCN cells grown from embryonic rats. He found that each cell had a distinct circadian rhythm, but in this artificial environment, the cells did not synchronize with each other. Each of these brain cells was, indeed, a miniature clock.

All in the genes?

We now have advanced techniques for working with genes—those DNA units, and their protein products, that contain the instructions that shape our bodies and how they work. We can identify which specific genes influence each of the functions of a cell. So, as the human internal clock lies within each cell, the program for timekeeping must be found in our genes.

The first clues as to the identity of "clock genes'" came from studying a variety of organisms whose circadian clocks ran fast, slow, or apparently not at all, and comparing them to those with normal rhythms. It was found that the circadian clock is based on a "feedback" principle. At its core, the clock consists of clock genes that, in turn, produce clock proteins. These proteins accumulate within the cell and gradually inhibit the activity of their own genes, either directly or through an intermediate set of genes. The proteins themselves are gradually broken down, which releases the clock genes from inhibition and so starts the cycle of clock protein production all over again.

This simple feedback mechanism appears to have a long evolutionary history, as it exists in all of the organisms that have been studied in depth so far. Also, very distantly related species share many of the same clock genes. There is still much to be learned about exactly how the circadian feedback loop is regulated, how it is affected by environmental stimuli such as light received by the eye, and how it controls the activities of the clock cells. However, it seems that the answer is not that far out of our grasp.

Internal clocks and the human condition

A great deal has been learned about the internal clocks that regulate our daily rhythms, but much remains to be discovered. We know where the clock is located in a mammal's brain, but our understanding of its internal mechanism, and how this is reset by external stimuli such as light and dark, is still incomplete. We do not know how the clock communicates with the rest of the brain and body to generate our complex rhythms of physiology and behavior. It is unclear whether the circadian clock provides signals that contribute to our sense of time in terms of hours and days.

What we do know is that circadian timing contributes directly to our health and performance. Jet lag, and changing from day to night work, produce such well-known effects as insomnia, fatigue, and digestive problems. Poor adaptation is associated with reduced on-the-job productivity and increased accidents. It is possible to use bright light to accelerate the process of shifting, but this may be difficult to implement in many work settings. We do not yet have effective drugs to help reset the clock, although the pineal hormone melatonin has shown some promise.

Seasonal, annual, and other clocks

Is the SCN the only biological clock? In at least some birds and reptiles, the retina and the brain's pineal gland also appear capable of some kind of internal cycle. The retina of mammals does contain an inbuilt twenty-four-hour mechanism, but this is important only for the retina's own rhythms (for example, timing the daily replacement of light-sensitive pigments). A mammal's pineal gland is not really a clock at all. It does secrete the hormone melatonin each night, but this daily rhythm is dependent on timing information from the SCN. Many animals exhibit seasonal rhythms in behavior and physiology, and in some these are controlled by an "annual clock." The ground squirrel gains weight each fall as the days become shorter, and spends much of the winter in hibernation. This seasonal cycle has been shown to persist for several years under controlled constant conditions of fixed day length, temperature, and food availability. So far, the brain site of this annual clock has eluded researchers. Humans also exhibit signs of seasonal rhythms, but whether these are truly inbuilt, persisting cycles comparable to a squirrel's hibernation rhythm has never been tested. The best example of a human annual rhythm is "seasonal affective disorder" (SAD), a type of depression that occurs during the short days of fall and winter and disappears in the summer. This can be successfully treated by exposure to bright light, suggesting that it is directly related to external seasonal fluctuations in sunlight, rather than an internal annual clock. Changes in human behavior corresponding to the "lunar month" (the twenty-nine and one half days between successive new moons) are well described in folklore, fiction, and anecdote; but there is no evidence that such rhythms, if real, correspond to the beating of an internal "lunar clock."

There is evidence that abnormal circadian timing causes certain sleep disorders, but for many individuals we do not yet have effective treatments. Disruptions of circadian timing are characteristic of certain psychiatric disorders, particularly depression, but we do not know which is the cause and which the effect. There are circadian times (primarily the early morning hours) when stroke and heart attack are more common, and there is a circadian variation in our responses to many therapeutic and recreational drugs. However, the principles of "chronopharmacology," which aims to make medicinal drugs most effective by administering them at certain times of day, have been slow to be embraced by the medical community.

It may be that circadian rhythms deteriorate with age, contributing to daytime sleepiness, nighttime insomnia, and possibly the ultimate failure of other physiological systems. We expect in the coming years to gain a great deal more insight into the clocks that time us, for the better of the human condition.

ome animals, such as the
ormouse, spend much of
e winter in hibernation.

3: prehistoric timekeepers

Clive Ruggles

Staying in tune with nature

in the southwest of England.

Midwinter spring is its own season…
Suspended in time between pole
and tropic T. S. Eliot

How time-conscious were prehistoric humans? We have no direct way of telling what went on in their minds. By definition they left no written evidence. Even in parts of the world where oral traditions, passed down over many generations, survived into historic times, it is difficult to determine when they arose and how much they have been altered along the way.

The earliest timekeepers

We do have the archaeological record to go on. The Neolithic and Bronze Age communities of northwestern Europe, for example, have left material remains that include tools, ornaments, and great stone monuments still conspicuous in today's landscape.

Years ago, it was considered impossible to go very far with this material evidence. The basic task of the archaeologist was to discover, identify, and classify the monuments and artifacts. It was possible to draw some conclusions about trade and exchange between communities, and at a stretch about the nature of those communities: how they were organized, what were the power structures, social hierarchies, spheres of influence, and structures of power.

Castlerigg stone circle, i
Yorkshire, England.

With the rise of "environmental archaeology" in the mid-twentieth century, artifacts such as bones, seeds, and mollusk shells yielded valuable information about people's health, diet, and the environment within which they lived. It was difficult, by comparison, to ask what people believed, how they understood the world and their place within it, or how they perceived space and time. This was a venture well beyond what experts thought the evidence had any power to inform us about.

In more recent times, ideas from anthropology, history, and other disciplines help archaeologists interpret more deeply. There are still no hard-and-fast answers about what prehistoric people thought and believed. But research that is fair with the evidence—research that does not simply select data that seem to support a favored idea while ignoring the rest—can at least move beyond pure speculation.

How intelligent were our ancestors?

The famous English stone circle, Stonehenge, was built by Neolithic people in the third millennium B.C., and has stood as a mute reminder of colossal human endeavor ever since. The puzzle of what exactly it was for has been an irresistible one. In the 1960s, British astronomer Fred Hoyle claimed that the structure was some sort of computer or calculating device. Similarly, the engineer Alexander Thom was sure that dozens of standing stone monuments in Scotland and elsewhere, erected during the Neolithic and Bronze Age, were sophisticated instruments for observing the sun and moon.

Again in the 1960s, the astronomer Gerald Hawkins popularized the idea that Stonehenge incorporated a great many deliberate solar and lunar alignments (see the diagram on the following page). The archaeologist Richard Atkinson, who had excavated at Stonehenge, vehemently attacked this idea. Atkinson at one point described the people who built Stonehenge as "howling barbarians," incapable of the scientific thinking Hawkins was crediting them with.

This remark was held up to ridicule by defenders of Hawkins, who took it to indicate that Atkinson failed to grasp that Neolithic people could be our intellectual equals. People in later prehistory, as far back as around forty thousand years ago—the beginning of the period known to archaeologists as the Upper Paleolithic—were biologically identical to us. There is no reason to think that they had inferior intelligence. Forty thousand years is virtually nothing in evolutionary terms.

Yet the builders of Stonehenge—though they had an intelligence broadly equivalent to anyone in the modern world—had no background of scientific thought and technological know-how. They didn't have the same shoulders to stand on as a modern stargazer—a line of mathematicians and astronomers reaching back two-and-one-half thousand years, who

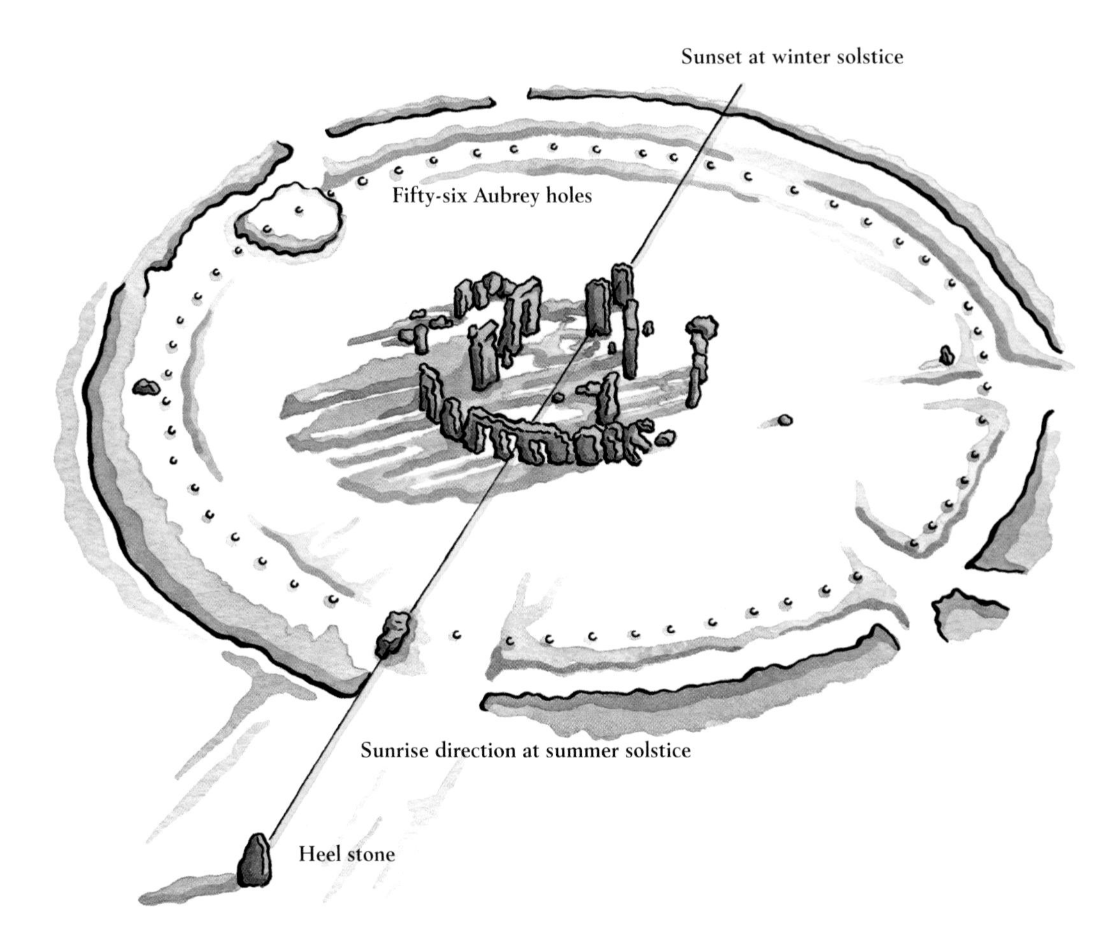

Above: Stonehenge appears to be roughly aligned on the midsummer sunrise and the midwinter sunset, but there is no independent evidence to confirm that the builders intended this. The fifty-six Aubrey holes—seen here as an outer ring of white dots—could have been used to predict eclipses, by moving markers around from hole to hole. But this is only if the people who used Stonehenge had sufficient astronomical knowledge, and if they were interested in predicting eclipses. However, there is no independent evidence that they did or that they were.

Below: The Abri Blanchard bone may be a thirty-thousand-year-old lunar calendar, but this interpretation of it has been questioned.

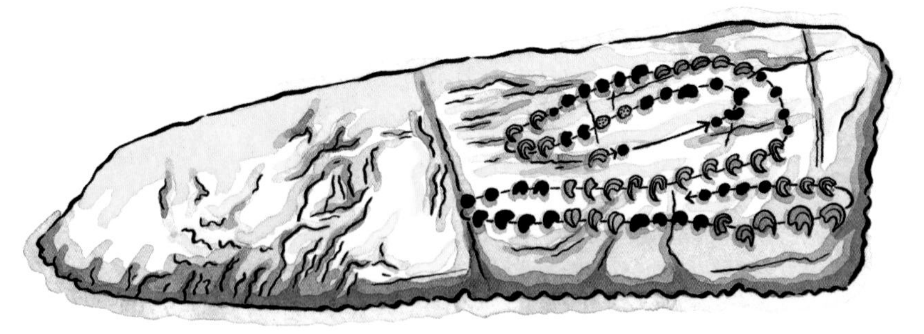

preserved and passed on their findings from generation to generation in increasingly sophisticated written documents.

Just as important is the fact that prehistoric thoughts and actions were unlikely to move along the same lines as our own. It is ethnocentric to assume that all intelligent people have the same motivations, concerns, and conceptual framework as us. See Chapter Five for more on the diverse range of ways people all over the world have conceptualized time outside the framework of modern Western thought.

The Abri Blanchard bone

Alexander Marshack, an expert in microscopic analysis at the Peabody Museum in Boston, has spent much of his career studying human incisions on prehistoric artifacts. One of the most fascinating of these is a small bone, part of an eagle's wing, originally found in a cave at Abri Blanchard in the Dordogne Valley in France. It was deposited there by hunters of the Upper Paleolithic period in around 30,000 B.C., at a time when much of Britain was covered in ice sheets. The period is known for its striking cave art, depicting animals such as deer, bison, and horses.

The Abri Blanchard bone, however, contains nothing pictorial, just a series of notches forming a winding line, bending back and forth in a serpentine pattern. The reason for its importance in archaeology is that, as Marshack has claimed, the marks were made not in one sitting, but using a number of different implements and different techniques. The obvious conclusion is that they represent a series of tally marks, counting off something.

This immediately raises various possibilities, such as recording animal kills, but Marshack contends instead that the bone is a rudimentary calendar, marking the passage of time using the most obvious regular cycle in the sky: the phase cycle of the moon, also known as the synodic month. The period from one new moon to the next is between twenty-nine and thirty days. According to Marshack, each notch on the Abri Blanchard bone represents a day, with the days of the waxing moon being marked off in one direction and those of the waning moon in the other. The line changes direction around the times of full and new moon, carrying on for about two months. If this is true, the bone would represent one of the earliest known devices for recording the passage of time.

A number of Marshack's assumptions about the marks on this bone can be questioned. Chief among them is the idea of where the direction of the marks turns. The turns are not sharp, and we need to make a judgment about where we consider that the change of direction actually occurs. If we make our choice solely on the basis that the numbers fit the lunar phase cycle, then this is not exactly being fair with the evidence. The ideal would be to find many similar artifacts and undertake a statistical analysis. Unfortunately, while many more engraved artifacts from the Upper Paleolithic do exist, and several have been studied by Marshack and also claimed to be lunar calendars, their designs are generally more complex, and there is even more capacity for judiciously selecting the data that best fit the theory. The question remains an open one.

The first monuments

A huge chronological gap separates the production of the engraved artifacts of the Upper Paleolithic from the time when we start to see more obvious evidence of people's perceptions of time. The key event is the appearance of monuments.

In around 8000 B.C., the ice sheets that had covered much of northern Europe for many tens of thousands of years finally receded as the temperatures rose. By 6000 B.C. Britain, for example, which two millennia earlier had been arctic tundra, was thickly covered in mixed deciduous forest. During this period, known as the Mesolithic, the predominant methods of subsistence were fishing, gathering, and hunting. Few clues (at least, few clues that have yet been recognized) about people's perceptions of time remain from the Mesolithic period.

Farming, which conventionally marks the onset of the Neolithic age, had already begun to spread into southeastern Europe from the Near East by this time, and it gradually spread northward and westward across Europe, reaching Britain around 5000 B.C. The transition to farming was a complex process, differing considerably from place to place. Herding and agriculture generally resulted in roving populations becoming sedentary, starting to live in fixed villages. It also resulted in the gradual clearance of tracts of natural forests, opening up the land for grazing, crops, and affording wider views.

It is in the Neolithic period that we see the appearance throughout much of western Europe, especially around the northern and western extremities, of considerable numbers of conspicuous monuments. These include many thousands of burial monuments of a variety of sizes and designs, built of different materials such as earth, timber, and stone. Certain traditions and conventions held sway in different regions and different times throughout the Neolithic and, in some places, well into the succeeding Bronze Age, i.e., well past 2000 B.C. It is these monuments that give us some of the most vital clues about how people were then perceiving time.

ecorated stones
rround the burial
ound at the Newgrange
mb in Ireland.

Death and the sun

One of the most impressive Neolithic monuments is the great passage tomb of Newgrange, part of a complex of tombs on the banks of the River Boyne in County Meath, Ireland. A huge mound some 264 feet in diameter covers a great stone chamber with three smaller recesses where, in common with many other Neolithic tombs, the bones of the dead (in this case cremated) appear to have been placed over a period of time, along with other grave goods. The chamber is accessed by a sixty-two-foot-long passage from an entrance on the southeast side.

Built a little before 3000 B.C., the Newgrange tomb is famous for the fact that a "roofbox" structure above the entrance admits the light of the sun for a few minutes after sunrise on days around the winter solstice (see diagram on following page). The sunlight would have shone along the entire passage, illuminating the tomb spectacularly for a few precious minutes before it was plunged back again into its customary darkness.

We can only speculate about what this meant to those who built and used the Newgrange tomb. It seems clear that this was not an "observatory" or "calendar" in the sense that people did not come here to tell whether the shortest day of the year had arrived. Rather, the spectacle confirmed existing knowledge, reemphasizing a belief about the connection between the dead and the sun, or light, or a particular time in the seasonal cycle—the time when the sun stopped withdrawing southward and began its life-giving journey northward to make a time when crops could grow and ripen in its warmth. This place of the ancestors was charged with special significance at certain rare times, occurring quite regularly in the seasonal cycle. We can imagine the penetration of light into the tomb being accompanied by other seasonal rituals and ceremonies, although we have no direct evidence for this. On the other hand, the fact that the sun entered not through the entrance itself but through the roof-box above it ensured that even when the entrance was eventually blocked off to the living, the event would still occur for the benefit of the ancestors.

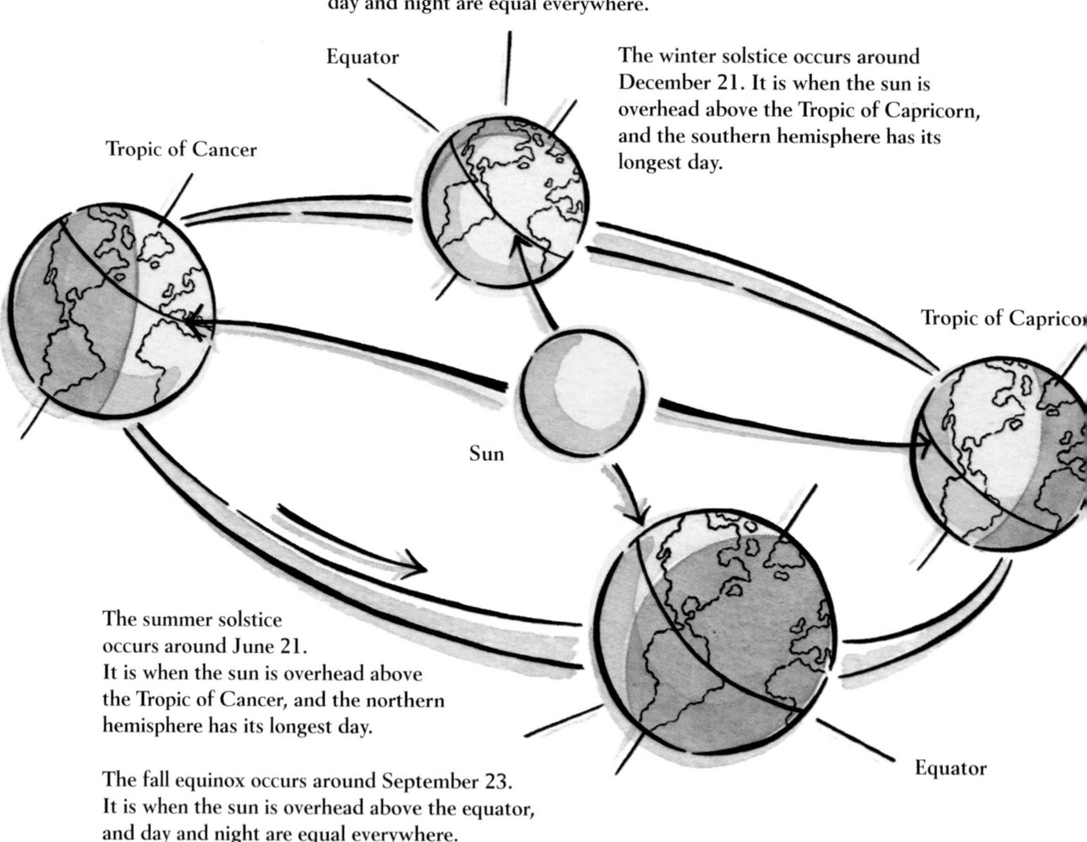

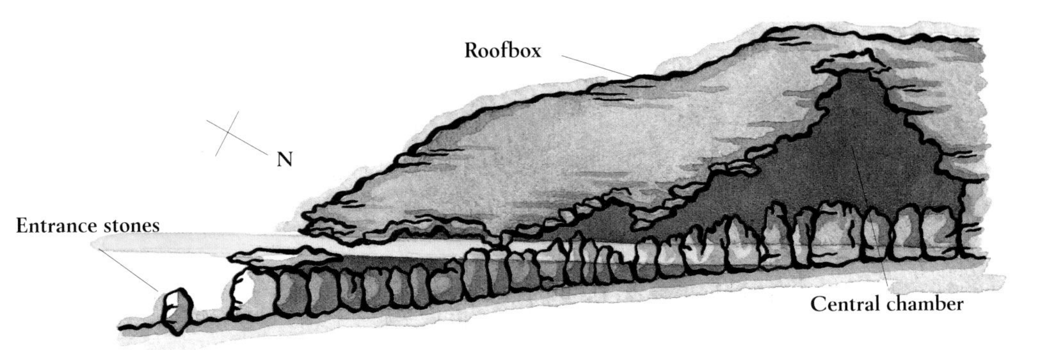

Crosssection of Newgrange tomb: at the winter solstice, and two weeks on either side, the sun's rays enter by the roofbox, a slit running on the top, the length of the 62-foot passage, and illuminates the 20-foot (6-meter) high burial chamber.

The orientations of Neolithic tombs

While a few other Neolithic tombs face the sunrise or sunset at a solstice, the great majority do not. This immediately raises the possibility that the alignment at Newgrange might have arisen without any intention on the part of the builders. Some archaeologists have argued that the Newgrange roofbox had some other purpose entirely, and the lighting of the burial chamber at the time of the solstice may have come about purely by chance.

Other archaeologists list several examples of tombs oriented not only toward sunrise or sunset at the solstices, but also at the equinoxes. These occur approximately halfway between the solstices, when the sun rises roughly due east and sets due west (see diagram left). But selecting examples of orientations upon targets that we might consider significant and ignoring the rest is not really a way of showing what prehistoric people intended to do. Research that examines whole groups of similar tombs, which represent local traditions in particular places at particular times, has a much better chance of probing the prehistoric mind.

Tomb patterns

During the past few years an archaeoastronomer from Cambridge University in England, Michael Hoskin, examined many hundreds of prehistoric tombs in different parts of Europe, measuring their orientation. His conclusion was that nearly all local groups of tombs conform to consistent patterns. The orientations within each pattern are restricted to just a portion of the compass—for example, between northeast and south. This in itself shows that orientation was of almost universal importance. It also shows that orientation was to the sky, since there is no other way to account for its consistency from a variety of locations. It is not possible, for example, that the orientation was to the direction of the prevailing wind, since the local topography would produce significant variations from place to place in many areas.

Facing the sun

According to the work of Michael Hoskin, local traditions of tomb orientation invariably conform to one of a small number of distinctive "signatures" which relate to the path of the sun through the sky at different times in the seasonal year.

The directions of sunrise and sunset on the longest and shortest days of the year define four directions in the northeast, southeast, northwest, and southwest quarters of the compass, the exact directions depending upon the latitude and the altitude of the horizon at a given place. These four directions divide the compass into four "quadrants" (not generally equal in extent) as shown in the diagram on pages 50 and 51.

There is a quite staggering consistency in patterns of orientation to the sun. Some groups of tombs consistently face points in the sun-rising segment of the horizon. Others face a wider range of points, including both the sun-rising and sun-climbing segments. Others are consistently oriented upon sun-falling and sun-setting places, and so on. There are only a very few examples (out of several thousand) of tombs facing north. It is not common for tombs to face, as Newgrange does, sunrise or sunset at the solstice itself, and there is no evidence—from tomb orientations—of a prehistoric interest in the equinoxes.

Hoskin has suggested that in some cases the orientation of a particular tomb was determined by the direction of sunrise or sunset on the day construction began. If this is the case, then the pattern of orientation of a group of tombs may tell us something of which season tomb construction preferentially took place.

There can be little doubt that the seasonal cycle of the sun was of the utmost importance when constructing tombs for ancestors, not just here and there, but as common practice widespread in many parts of Europe over several millennia of later prehistory. The extent to which tombs may have incorporated more precise calendrical indications in certain places and certain times remains an open question.

Circular temples and timed ceremonies

The Neolithic period saw the widespread construction not only of tombs but also of other monuments. In Britain, for example, many hundreds of circular ditch-and-bank enclosures known as henges and stone circles remain a spectacular part of today's landscape. Recently, traces of numerous timber circles dating to a similar period have been found, although the timber posts themselves have long since rotted away. Earthen, timber, and stone circles were often associated with each other. Different types of circle often succeeded each other as sites were modified and reused.

A circle is the most efficient way to mark off a segment of land for ceremonial use, but there may have been considerably more to it than that. Anthropologists have revealed the importance of central places for small communities—places that are perceived to be at the very center of the world. Archaeologists such as Richard Bradley have explored the possibility that circles acted as microcosms, reflecting the properties of the wider world. This might result, for example, in taller stones being placed so that they line up with mountains on the horizon, henge entrances being aligned with water sources, or outlying stones and posts being aligned with significant and recurring celestial events.

All this would have been part of ensuring that what was perceived as the most important of all places would be fully in tune with the natural world. Archaeologists test ideas such as these by taking sample groups of monuments and looking systematically at their place within the landscape. For example, the archaeologist Aubrey Burl has shown that a local group of large stone circles in Cumbria incorporate preferential alignments upon the cardinal compass directions and the directions of sunrise and sunset at the solstices. This is in tune with Hoskin's conclusions for tombs.

One distinctive group of stone circles may give us stronger clues about the possible rituals that took place there. These are the so-called "recumbent stone circles" of the Grampian region of eastern Scotland. They all have a single, large stone that has been placed on its side (the "recumbent stone"). Always oriented roughly within the quarter of the horizon centered upon south-by-southwest, they exhibit a clear pattern of orientation relating to the moon. Excavated sites show scatterings of fragments of quartz—a white stone that in some human communities actually symbolizes moonlight—in front of the recumbent stone. This is an intriguing hint that ceremonies took place here that celebrated the moon and were regulated by its cycles.

A time of pilgrimage

Not all nonfunerary monuments were circular. Other common Neolithic and Bronze Age monuments are linear in design. The earliest of these are the so-called "cursus" monuments—earthworks consisting of two parallel ditches and banks, generally several tens of meters apart. They date from the fourth millennium B.C. Most are only detectable as crop marks on aerial photographs, but many dozens of examples are now known, mostly in southern England. The longest is the Dorset cursus, which runs for some six miles across Cranborne Chase.

Cursus monuments are enigmatic, but seem to have been used in some kind of ceremonial or ritualized movement of people through the landscape. A complex of four cursus monuments at Rudston in Yorkshire

Megalithic alignments

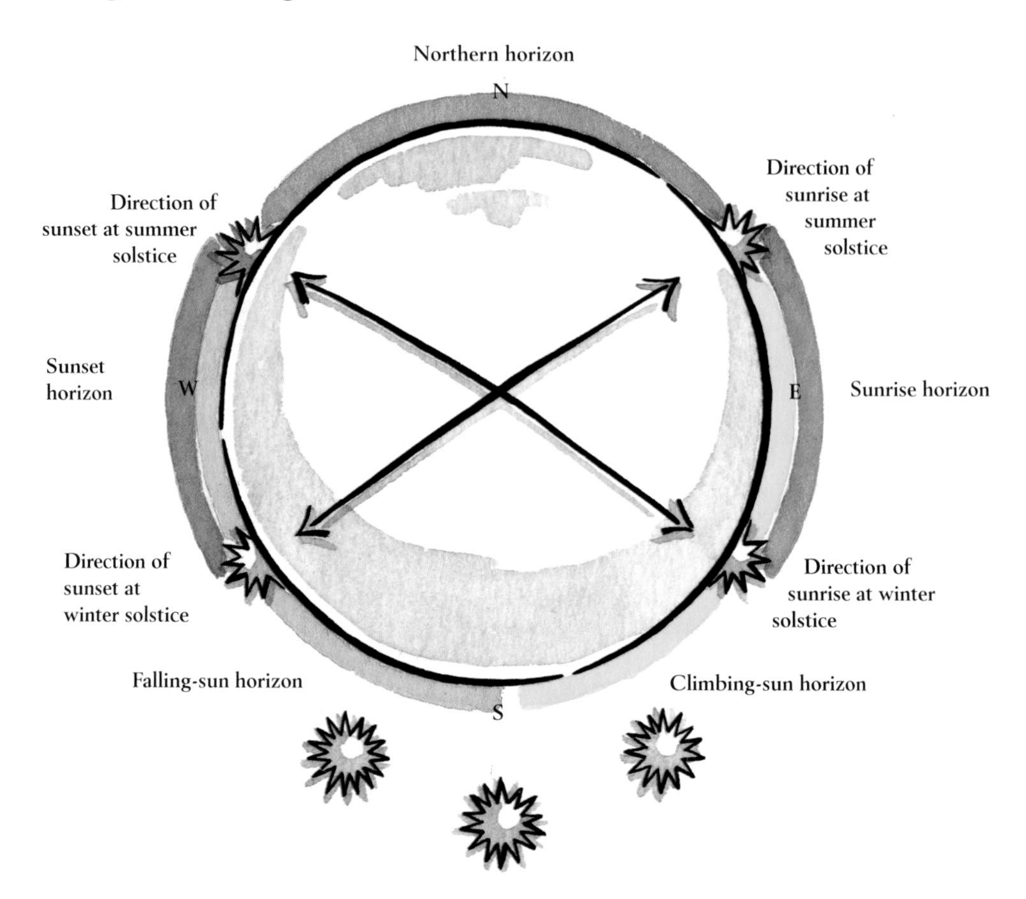

Above: The directions of sunrise and sunset at the solstices divide the compass into four. The sun rises on the horizon of the eastern segment throughout the year and sets on the horizon of the western segment. During the day, in the northern hemisphere, the sun can be seen above the southern horizon, climbing in the sky while it is to the east of due south, and falling when it is to the west. In the northern hemisphere, the sun never appears above the northern horizon. *Right:* The two most common sun-facing patterns for megalithic tombs. *Above right* are illustrated the orientations of forty-one tombs at Montefrio, west of Granada in Spain, exemplifying the sunrise pattern, with thirty-eight of the tombs orientated towards the sunrise segment of the horizon.

Below right: The orientations of thirty-four tombs on the Rio Gor, east of Granada, exemplifying the pattern in which tombs appear to be orientated toward the climbing sun. Of more than two thousand tombs investigated in the western Mediterranean islands, Italy, southern France, Spain, and Portugal, about 95 percent are orientated either to the sunrise, the climbing sun, or both. In the Languedoc region of southern France, on the Balearic islands and in one tiny region of Catalonia in Spain, a west-facing pattern is prominent.

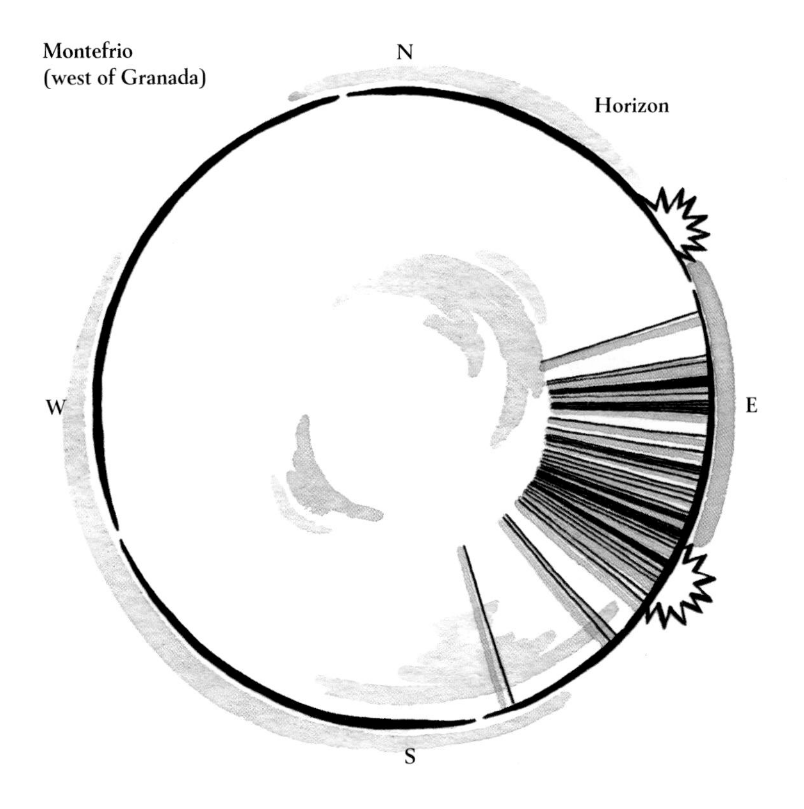
Montefrio
(west of Granada)
N
Horizon
W
E
S

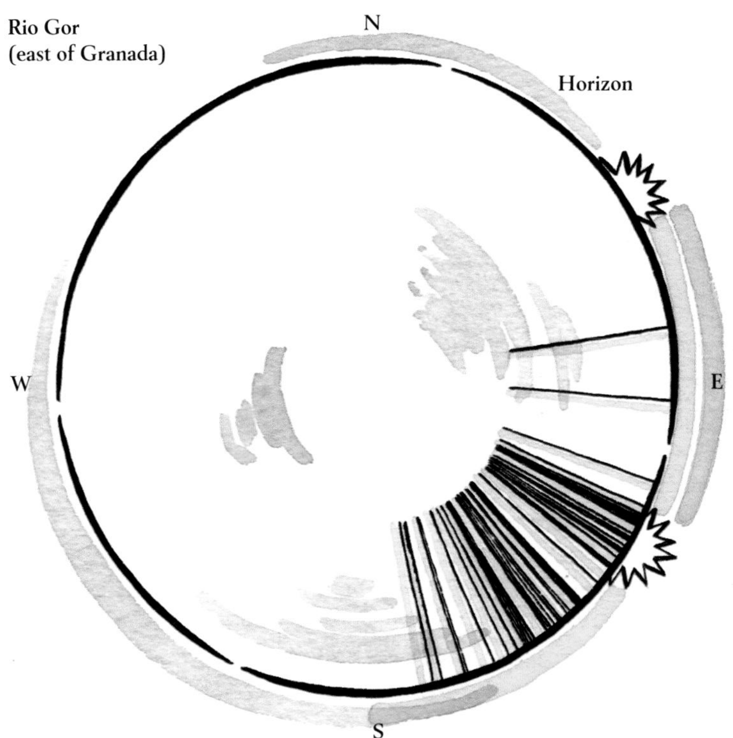
Rio Gor
(east of Granada)
N
Horizon
W
E
S

converge on a turning point in a major river valley from various high places in the surrounding landscape. Many cursuses connect one high place to another.

Did this ceremonial movement take place at special times of the year? It may not be coincidental that the longest-known, the Dorset cursus, is aligned on sunset at the winter solstice. The Dorchester cursus in Oxfordshire, England is aligned with sunset at the summer solstice. One could easily imagine the sight of a sunrise or sunset either starting or ending a ceremonial pilgrimage along one of these cursuses. The problem is that the orientation of these two cursuses seems to be exceptional; many other factors seem to have determined the orientation of cursuses in other places, and so the question remains open.

Rows and avenues of standing stones started to appear later in the Neolithic period. Short rows of between three and six stones became ubiquitous in several parts of Scotland and Ireland by the Early Bronze Age. Statistical studies of their orientations have shown that the directions in which they point are often associated with moonrise or moonset. In addition, intensive studies on the Isle of Mull have shown that a cluster of short stone rows there also served to mark places where a prominent distant mountain came into view. This might imply that such monuments marked significant places on pathways through the landscape. It is possible that routes were trodden ceremonially, perhaps at auspicious times regulated by the observations of the moon.

Megalithic observatories?

In three books published between 1967 and 1978, Alexander Thom published studies of hundreds of stone monuments in Britain, particularly in western Scotland. His claim was that they were used as instruments to observe the motions of the sun and moon with great precision. The key to this precision was to use features such as notches in distant mountainous horizons to pinpoint changes in rising or setting positions. Standing stones simply marked where to stand and where to look. Thom also claimed that prehistoric people had a precisely defined unit of measurement, the "megalithic yard," and that monuments such as stone rings were laid out using complex geometrical constructions involving, for example, Pythagorean triangles.

Thom's conclusions were backed up by a formal statistical analysis, but reassessments eventually showed that it was flawed by subtle, no doubt unconscious, biases in the selection of data. For example, there might be several horizon features within a stretch of mountainous horizon that might have been "indicated" by an alignment of standing stones, but only one of these corresponded to a rising or setting position of the moon that Thom believed significant. The other features were then omitted from his statistical analysis. Similar conclusions have now been reached with regard to Thom's ideas on measurement and geometry.

There really is no evidence to support high-precision alignments. On the other hand, the reassessments did not rule out the possibility of astronomical alignments of a lower precision, and indeed Thom's work helped to bring attention to some of these in the first place.

Ritual observance

The concept of ancient monuments functioning as astronomical instruments, or "observatories," has received a great deal of attention over the years. Yet archaeologists now feel that while astronomical alignments were deliberately incorporated into some of these monuments, they almost certainly reflected existing knowledge. For example, they symbolized the integration of a ritual monument with the functioning of nature, rather than serving as astronomical or calendrical "tools." There are simpler ways of telling the time of year than building a henge—for example, by observing the changing stars and constellations visible in the evening sky just after sunset, or the morning sky just before dawn (see Chapter Four). Even if the motions of the sun or moon were tracked from a fixed spot against features on the natural horizon, it would not be necessary to use anything but temporary markers to identify the spot.

Although Alexander Thom is now thought to have been wrong in attributing great mathematical and astronomical sophistication (in our terms) to people in later prehistory, he was very much ahead of his time in thinking about the design of monuments in relation to the wider landscape. In the 1970s, when Thom's theories were causing a furious controversy between archaeologists and astronomers, archaeologists were inclined to believe that patterns of human activity in the landscape during later prehistory could be explained through ecology and economics. Yet, drawing from anthropology, most archaeologists now believe that we cannot ignore systems of belief.

The ways in which people conceptualize the world, and attempt to keep their actions in tune with the cosmos as they perceive it, are almost always significant factors in determining what it is that people do and where they do it. Cosmologies—systems of thought about the world—cannot be ignored if one is attempting to explain patterns of human activity in the landscape.

Most archaeologists now believe that we should view astronomical alignments as reflecting cosmologies. Anthropology shows us that people tend to make reference to cosmology in a variety of aspects of daily life. The correct timing of human actions has a ritual as well as a pragmatic aspect. Very often the one reinforces the other—seasonal ceremonies help to regulate an agricultural calendar. No clean separation exists between ritual activity and the mundane.

Iron age roundhouses, li this Iberian example, als show signs of being orientated eastward to tl rising sun.

This being the case, we would not expect cosmological symbolism to be confined to monuments—it would be found also in everyday dwelling places. Remains of dwelling places from the Neolithic and the earlier part of the Bronze Age are very scarce. One example is the Neolithic village of Barnhouse on Orkney, off the north coast of Scotland, which was excavated in the 1990s by Colin Richards. He found that the orientations of the houses were clearly related to the directions of the compass and to the solstices. As we move into the Iron Age, evidence becomes much more abundant in the archaeological record in the form of roundhouses (at the same time, large, communal monuments disappear). The roundhouses display clear patterns of easterly orientation relating to the sun. The study of "domestic cosmology" is in its infancy, but may have a great deal to tell us about perceptions of time.

Keeping in tune

It is unlikely that any prehistoric community perceived of time in abstract, as we do. To us, time is like a line marked off at regular intervals. We are seldom far from a calendar, radio, or clock that constantly monitors our steady progress along that line. Time provides a frame of reference—a backdrop—against which we live our lives. But in the absence of such aids it is much more likely that prehistoric people, in keeping with indigenous communities in history and around the world today, would have been aware of the passing of time by keeping their own actions in tune with various regularly recurring events going on in the natural world around them.

Many archaeologists now believe that prehistoric communities are likely to have perceived the landscape as a network of places charged with meaning. The construction of Neolithic monuments helped to structure the landscape and to add additional meaning, creating central places and ceremonial pathways. To us, space and time are separate abstract entities but to someone watching the winter sunrise from a Neolithic stone circle, they were bound together. The rising of the sun along an alignment may simply have defined the appropriate moment to be in that particular place performing a certain ritual. By doing so, the performers of the ritual ensured that the universe would remain in harmony for a while longer.

While such a scenario is speculative, it is almost certainly closer to the truth than the idea that people meticulously observed the solstices and equinoxes. Indeed, the assumption that half-way points between the solstices (the equinoxes) are significant only seems natural to us because we do perceive time and space as abstractions. Although the solstices were directly observable as the limits of the sun's motion, the concept of the equinox was almost certainly without meaning in prehistory. Instead, auspicious days in the ceremonial calendar are likely to have been determined by factors specific to particular places, such as significant events in their agricultural year or the day when, as viewed from the stone circle at the center of the universe, the sun is seen to rise behind a sacred mountain.

TYPUS SELENOSCOPICUS
LUNÆ PHASES
OS ADVARI
PHASES LVNÆ ACCRESCENTIS
PHASES LVNÆ DECRESCENTIS
LVNA COR NICVLARIS
LVNA SOLI CONIVNCTA
IN ASPECTV SEXTILI
LVNA DIMIDIATA
IN ASPECTV QVADRATO
LVNA IN ORBEM INSINVATA
IN ASPECTV TRINO VEL TRIGONO
LVNA SOLI OPPOSITA
PLENILVNIVM
NOMINA PHASIVM ET ASPECTVVM LUNÆ

4: the moon, the sun, and the stars

Robert Hannah

Counting the days and the years

The phases of the moon, as depicted in a hand-colored engraving from *The Celestial Atlas* of 1660.

When Zeus has brought to completion sixty more winter days, the star Arcturus, leaving behind the sacred stream of the ocean, first begins to rise and shine at the edges of evening Hesiod

The calendar adopted under Pope Gregory XIII more than four hundred years ago and now used everywhere in the world is the culmination of a sequence of developments in timekeeping that stretches back into prehistory. The earliest development was probably a recognition of the phases of the moon (see diagram below). After the cycle of day and night, the most easily observed regular change in the sky is the repeated progress of the moon from a thin crescent to a full moon, and then to a thin crescent before disappearing. The days of one lunar cycle are an obvious way of counting days off into groups, and as we saw in Chapter Three, markings on a bone fragment dating to the Upper Palaeolithic period—about 30,000 B.C.—may represent a tally of days grouped by a lunar cycle. The recorded history of the Gregorian calendar begins at a point when this way of grouping days still dominates timekeeping.

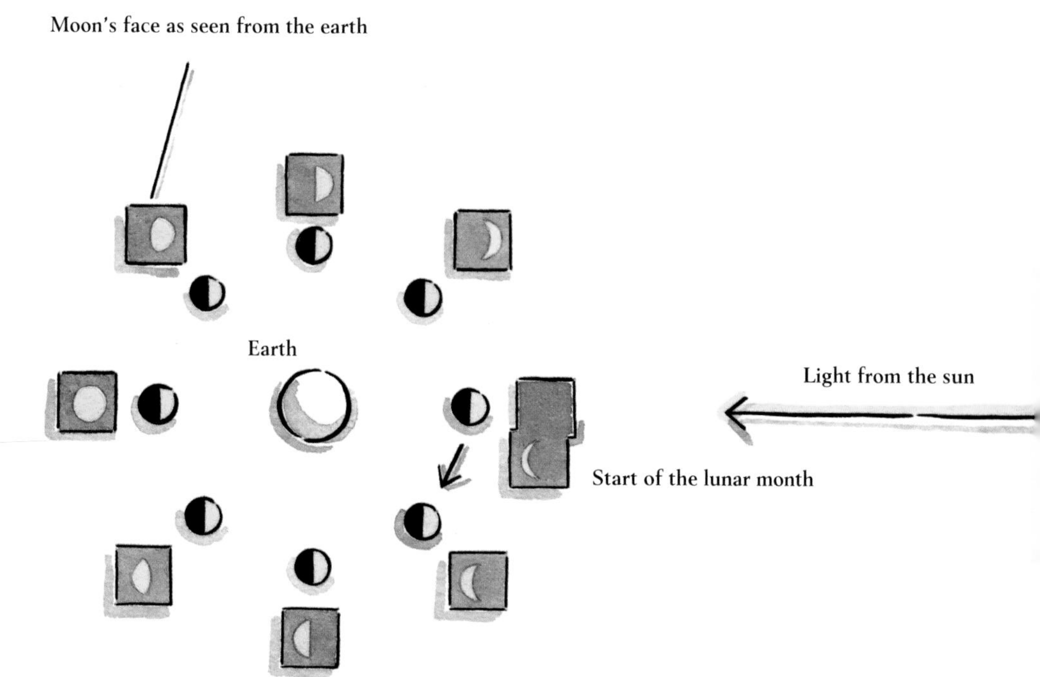

The phases of the moon

The Babylonian calendar

The ancient ancestor of the Gregorian calendar is the one used in Babylon more than three thousand five hundred years ago. It had twelve twenty-nine- or thirty-day months. The months were lunar, each beginning on the evening when the crescent of the new moon was first visible. The day started in the evening, at sunset (this practice survives to the present day, through a common heritage, in the timekeeping of Judaism and Islam).

The names of the months depended on the region, but those from Nippur (southeast of Baghdad in Iraq) came to dominate, first in Babylonia, and then in Assyria. These names are reflected in the months of the Jewish calendar, because of the Babylonian annexation of Jerusalem in 586 B.C. The series began at the spring equinox.

An extra month was added ("intercalated") at irregular intervals to bring the lunar calendar into line with the seasons. From the second millennium B.C. only the sixth and twelfth months (*Ululu* and *Addaru*) were repeated to produce this intercalation. The decision to repeat a month came from the king. Hammurabi, King of Babylon (1848–06 B.C.), decreed:

> *Tell Sin-iddinam, Hammurabi sends you the following message, "This year has an additional month. The coming month should be designated as the second month* Ululu, *and wherever the annual tax had been ordered to be brought in to Babylon on the 24th of the month* Tashritu *it should now be brought to Babylon on the 24th of the second month* Ululu.*"* (trans. J. Britton and C. Walker)

Babylonian and Jewish months

Babylonian	Jewish	Seasonal Point
Nisannu	Nisan	Spring equinox
Ayaru	Iyyar	
Simanu	Sivan	
Du'uzu	Tammuz	About summer solstice
Abu	Av	
Ululu	Elul	
Tashritu	Tishri	About autumn equinox
Arahsamnu	Heshvan	
Kislimu	Kislev	
Tebetu	Teveth	About winter solstice
Shabatu	Shevat	
Addaru	Adar	

The lunar calendar of Islam

The religious calendar of Islam is a lunar calendar, using the moon as its basis. Each month starts when the first sliver of the crescent moon is sighted, about two days after the actual new moon, which is invisible. The lunar month–the time from one new moon to the next–is not a whole number of days, but instead about twenty-nine-and-a-half days on average. To allow for this, the months of the Islamic year are alternately twenty-nine and thirty days in length.

A seasonal year (one organized according to the sun rather than the moon) consists of about 365.25 days. A purely lunar calendar cannot align easily with it–twelve lunar months add up to only 354 days, and so fall short of a seasonal year by about eleven days, while thirteen lunar months overshoot the seasonal year by more than eighteen days each year. The effect of this discrepancy between the lunar and solar years is that the Islamic religious year drifts through the seasons. For instance, the Islamic New Year will, over a period of solar years, run through each of the seasons, at one time occurring in winter, but in summer seventeen years later.

Ancient Arab zodiac by an unnamed artist, from *Astronomie populaire*.

The Islamic New Year (Muharram 1)

Gregorian dates 1994–2004

1994	June 11
1995	May 31
1996	May 20
1997	May 9
1998	April 28
1999	April 17
2000	April 6
2001	March 26
2002	March 15
2003	March 5
2004	February 22

The Islamic Months

Gregorian dates November 2000–February 2002

Ramadan (30 days)	November 28–December 27, 2000
Shawwal (29 days)	December 28, 2000–January 25, 2001
Dhu al-Qa'da (30 days)	January 26–February 24, 2001
Dhu al-Hijja (29 days)	February 25–March 25, 2001
Muharram (30 days)	March 26–April 24, 2001
Safar (29 days)	April 25–May 23, 2001
Rabi I (30 days)	May 24–June 22, 2001
Rabi II (29 days)	June 23–July 21, 2001
Jumada I (30 days)	July 22–August 20, 2001
Jumada II (29 days)	August 21–September 18, 2001
Rajab (30 days)	September 19–October 18, 2001
Sha'ban (29 days)	October 19–November 16, 2001
Ramadan (30 days)	November 17–December 16, 2001
Shawwal (29 days)	December 17, 2001–January 14, 2002
Dhu al-Qa'da (30 days)	January 15–February 13, 2002

From about 500 B.C. a consistent cycle of seven intercalations of an extra month over nineteen years was used. The sun and moon, it was understood, were brought back into the same positions that they held nineteen years earlier, and would do so again every nineteen years, because nineteen solar years equaled a complete number (235) of lunar months, each amounting to practically 6,940 days. Such a nineteen-year cycle is usually named "Metonic" after its presumed discoverer, the late-fifth century B.C. Greek astronomer Meton, but it is now considered more likely that the cycle was known in Babylonia before him, at the start of that same century.

The star calendar

Another means of reckoning the time by celestial bodies is through observation of the stars. The earth takes only twenty-three hours and fifty-six minutes to rotate on its axis. A day lasts four minutes longer than this because we have moved four minutes farther along our orbit around the sun before it reappears overhead at noon each day. This means that (by our twenty-four-hour solar clock) the rising and setting times for stars are four minutes earlier each night.

When a star rises during the daytime it is invisible because of the sun's light, but eventually, on one day of the year, it will rise in the east early enough to become briefly visible just ahead of sunrise. This first morning rising (also called "heliacal rising") of a selected star or star group makes a convenient time-marker. It is likely to be seen as a particularly significant time—a dramatic entrance—if you view the heavens as an annual procession of divine personalities coming into view and passing out of view.

Another notable day in a star's annual progression is its last evening rising—it will rise progressively earlier through the night until eventually it rises in the east just after the sun has set in the west at the beginning of the night. The star's first morning setting (when it sets in the west just before dawn in the east) and its last evening setting (when it sets in the west just before sunset) are also ready time signals.

Many ancient societies became aware of these phenomena. At least as early as 1000 B.C., the Babylonians included in their calendar horizon observations of the dawn and dusk risings of certain stars and dates for solstices and equinoxes.

The Metonic cycle

One solar year = 365.24219 days,
One lunar month = 29.53059 days,
which means that on average 12.3683 lunar months occur in each solar year.

In practice with this cycle, each year has to comprise a whole number of months, so some balance of twelve and thirteen months is likely to be needed to approximate a solar year on average over a period of time.

Nineteen solar years = almost 6,940 days
(actually 6,939.60161 days) and 235 lunar months = almost 6,940 days (actually 6,939.68865 days).

Therefore, nineteen solar years and 235 lunar months equal very nearly the same number of days.

So, over a period of nineteen years there will be almost exactly 235 complete months, which may be divided up as follows:

Twelve years of twelve months = 144 months
Seven years of thirteen months = 91 months
Nineteen years = 235 months

It is not known precisely which seven years of the nineteen-year cycle in Babylon had thirteen months instead of twelve.

A Babylonian star calendar

The seventh century B.C. compilation of earlier star catalogs called mul.apin ("The Plough"), after its opening line, gives an indication of how star calendars worked: extracts from it are quoted beside the star diagrams on the next few pages. In square brackets are the modern names of the stars and star groups the Babylonian text is thought to refer to. By each diagram is the modern equivalent of the time at which the constellation would have been visible.

This calendar is based on an ideal 360-day year, made up of twelve months, each of thirty days. No allowance is mentioned for the extra five or six days required to keep in line with the seasonal year of 365 or 366 days, but the various risings and settings will not recur from one year to the next on those same days without such an allowance being made. The identity of the constellations and stars observed is still a matter of some conjecture, but those named in square brackets represent current scholarly thinking. Another part of mul.apin has a different way of saying when the risings of key stars will occur, by day intervals divorced from any notion of the month:

> "Thirty-five days pass from the rising of the Fish to the rising of the Crook.
> Ten days pass from the rising of the Crook to the rising of the Stars.
> Twenty days pass from the rising of the Stars to the rising of the Bull of Heaven."

The total number of days in the list from beginning to end of the sequence remains 360.

How to read a Babylonian star calendar

The modern constellation name is indicated and the dark line is the horizon, below which the constellation is not yet visible.

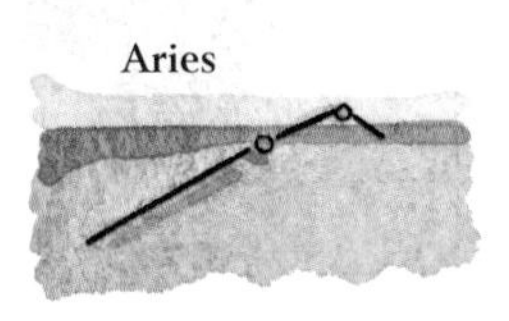

"On the 1st of Nisannu [spring equinox] the Hired Man [Aries] becomes visible"

Looking east from Babylon on March 31, 1000 B.C. at 5:15 A.M., just before sunrise

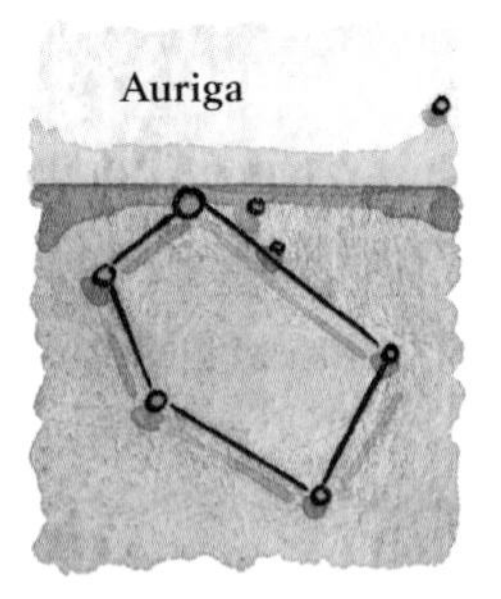

"On the 20th of Nisannu the Crook [Auriga] becomes visible"

Looking northeast from Babylon on April 20, 1000 B.C. at 4:49 A.M.

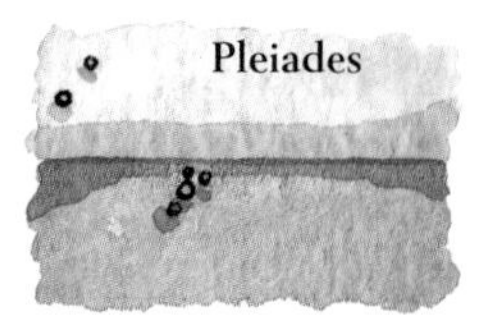

"On the 1st of Ayaru
the Stars [Pleiades] become visible"

Looking east by northeast from Babylon on April 30, 1000 B.C. at 4:36 A.M.

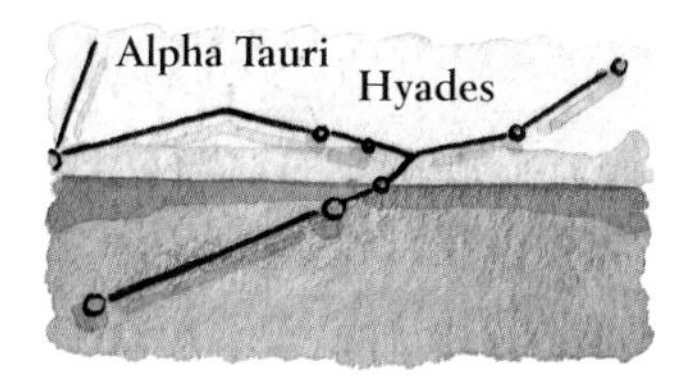

"On the 20th of Ayaru the Jaw of the Bull [Alpha Tauri and Hyades] becomes visible"

Looking east from Babylon on May 20, 1000 B.C. at 4:25 A.M.

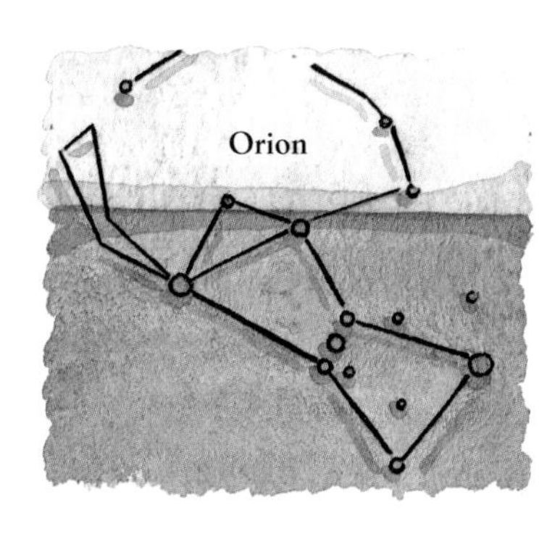

"On the 10th of Simanu the True Shepherd of Anu [Orion] and the Great Twins [Alpha and Beta Geminorum] become visible"

Looking east from Babylon on June 9, 1000 B.C. at 4:21 A.M.

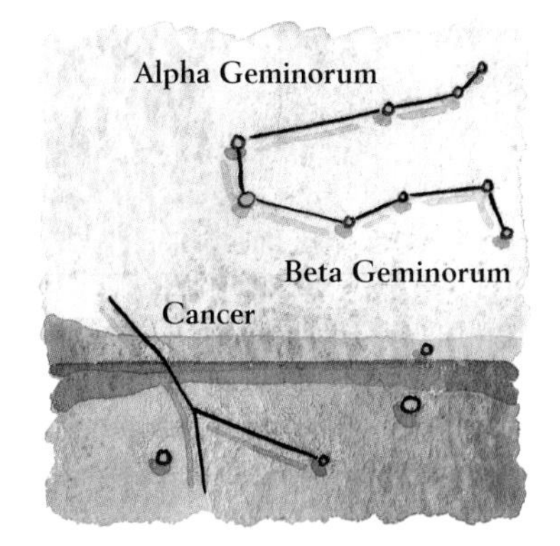

"On the 5th of Du'uzu [summer solstice] the Little Twins [among the Gemini] and the Crab [Cancer] become visible"

Looking east by northeast from Babylon on July 4, 1000 B.C. at 4:17 A.M.

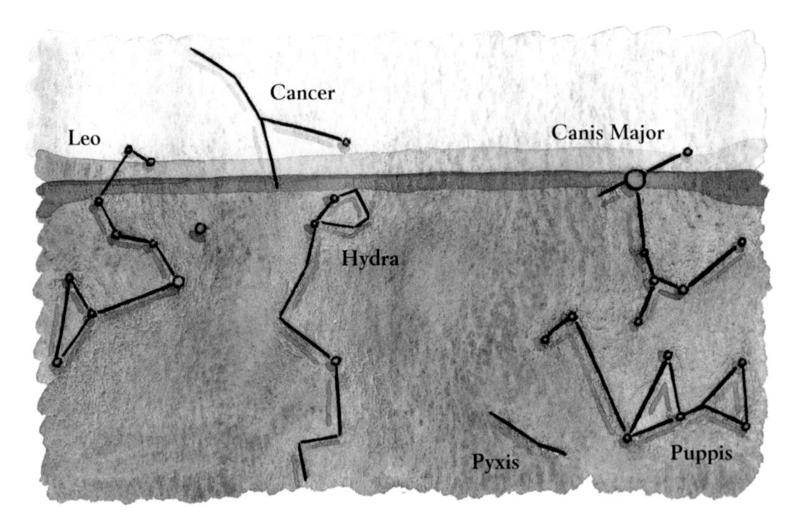

"On the 15th of Du'uzu the Arrow [including Canis Major and parts of Puppis and Pyxis], the Snake [Hydra] and the Lion [Leo] become visible"

Looking east from Babylon on July 14, 1000 B.C. at 4:31 A.M.

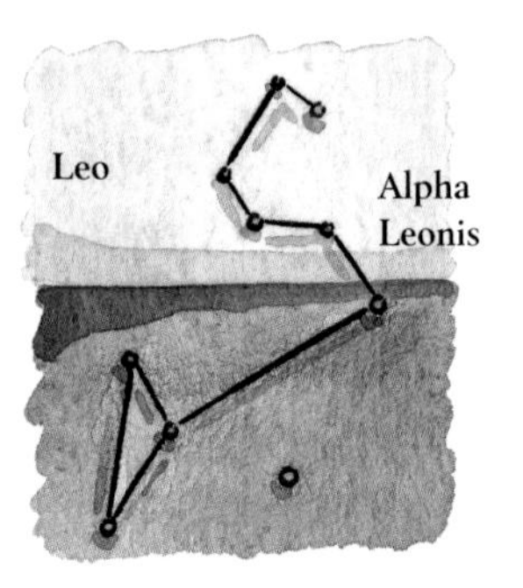

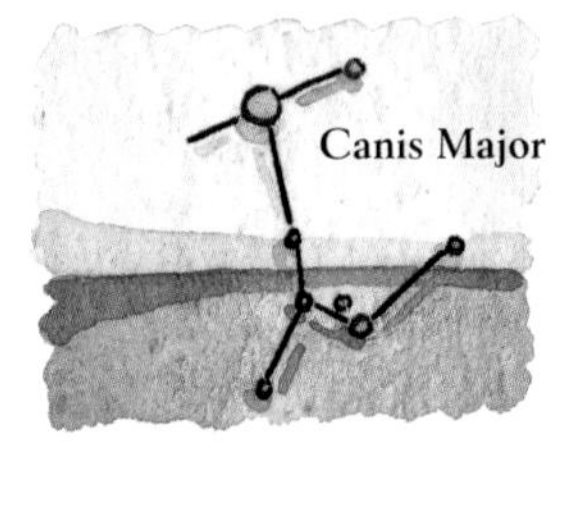

Far left and left: "On the 5th of Abu the Bow [including parts of Canis Major] and the King [Alpha Leonis] become visible"

Looking northeast (*above*) and southeast (*above left*) from Babylon on August 3, 1000 B.C. at 4:08 A.M.

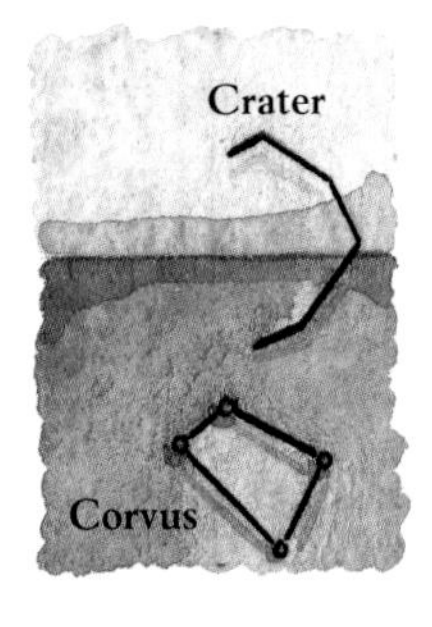

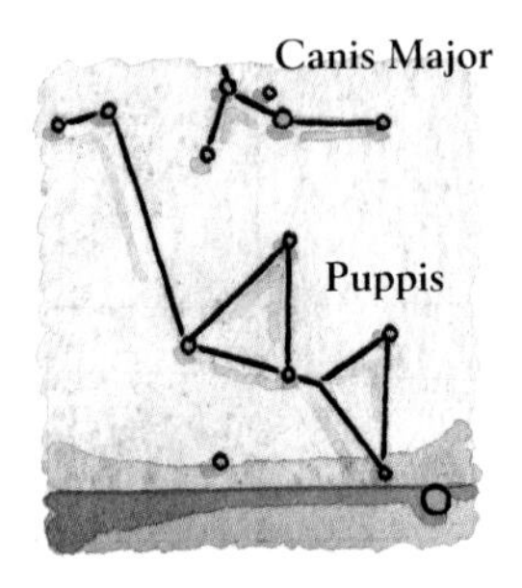

"On the 10th of Ululu the star of Eridu [in Puppis] and the Raven [Corvus and Crater] become visible"

Looking east (*above*) and southeast (*above right*) from Babylon on September 7, 1000 B.C. at 4:43 A.M.

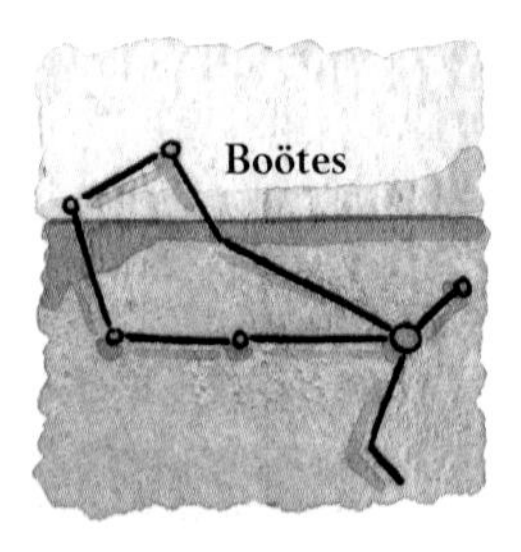

"On the 15th of Ululu the *supa* [Boötes], Enlil, becomes visible"

Looking north by northeast from Babylon on September 12, 1000 B.C. at 4:45 A.M.

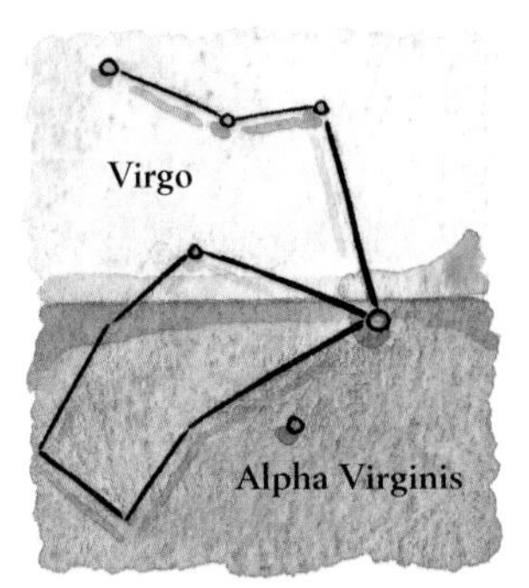

"On the 25th of Ululu the Furrow [Alpha Virginis] becomes visible"

Looking east from Babylon on September 22, 1000 B.C. at 5:12 A.M.

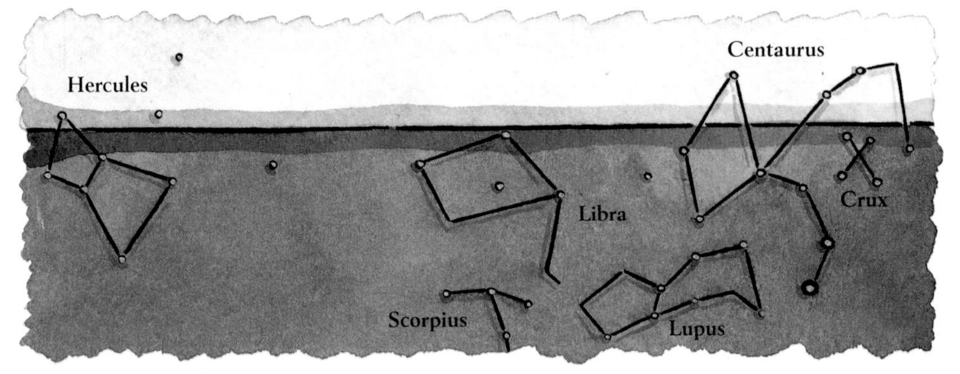

"On the 15th of Tashritu [about autumn equinox] the Scales [Libra and part of Virgo], the Mad Dog [Lupus and part of Scorpius], *en.te.na.bar.hum* [most of Centaurus and, probably, Crux], and the Dog [southern part of Hercules] become visible"

Looking east from Babylon on October 12, 1000 B.C. at 5:31 A.M.

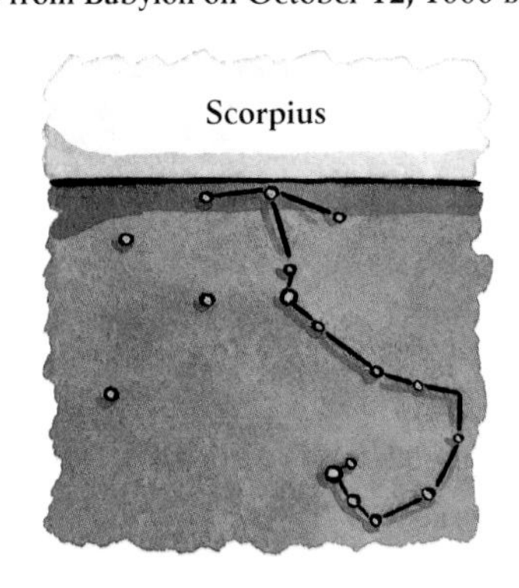

"On the 5th of Arahsamnu the Scorpion [Scorpius] becomes visible"

Looking east from Babylon on November 1, 1000 B.C. at 5:37 A.M.

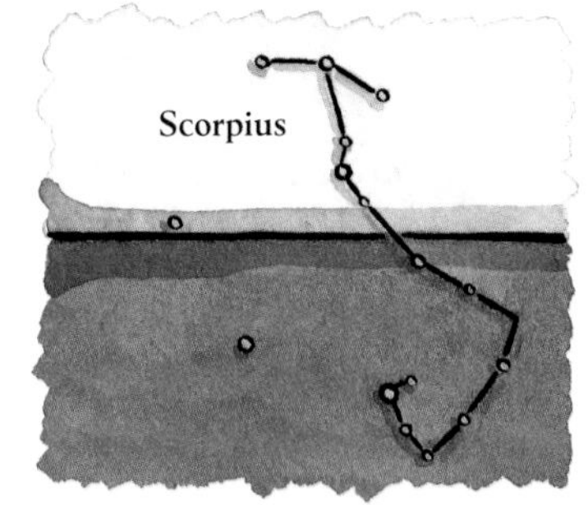

"On the 15th of Arahsamnu the She-Goat [Lyra] and the Breast of the Scorpion [in Scorpius] become visible"

Looking northeast (*far left*) and east by southeast (*left*) from Babylon on November 11, 1000 B.C. at 6:04 A.M.

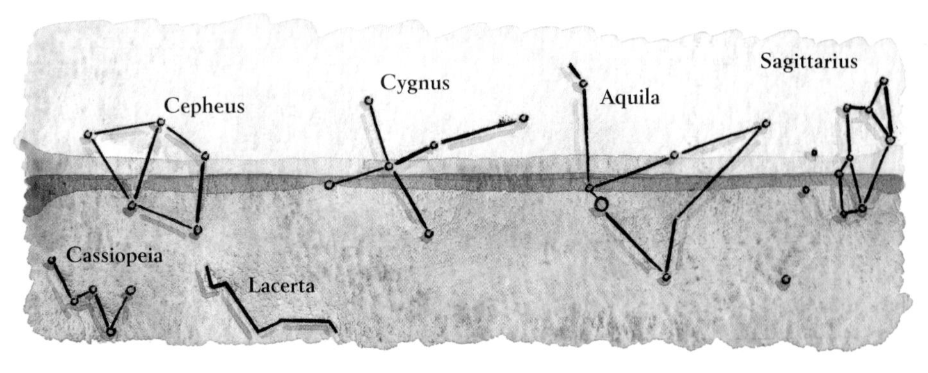

"On the 15th of Kislimu the Panther [Cygnus, Lacerta, and parts of Cassiopeia and Cepheus], the Eagle [most of Aquila], and Pabilsag [including Sagittarius] become visible"

Looking east by northeast from Babylon on December 11, 1000 B.C. at 6:19 A.M.

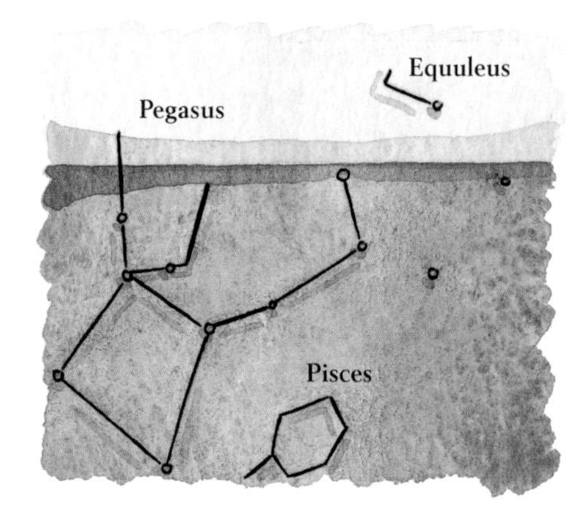

"On the 15th of Tebetu [winter solstice] the Swallow [including parts of Pegasus and Equuleus]...becomes visible in the east, and the Arrow [including Canis Major and parts of Puppis and Pyxis] becomes visible in the evening"

Looking east from Babylon on January 10, 999 B.C. at 6:36 A.M. (*left*) and looking southeast from Babylon on the same day at 6:37 P.M. (*below left*)

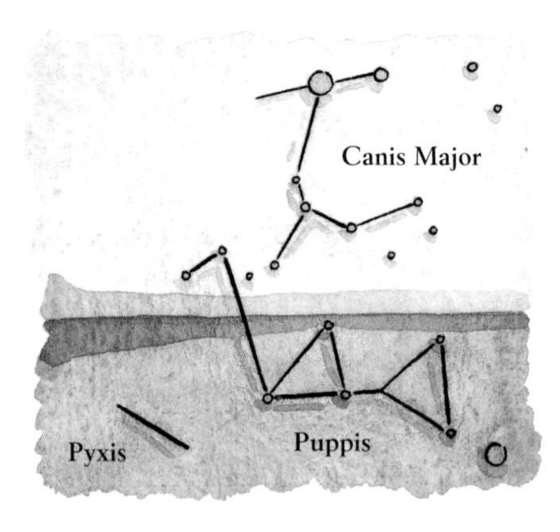

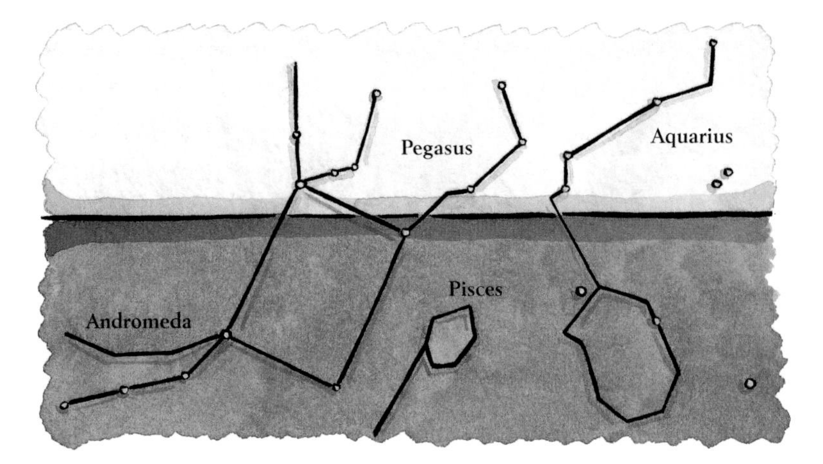

"On the 5th of Shabatu the Great One [Aquarius], the Field [including parts of Pegasus and Andromeda], and the Stag [eastern part of Andromeda] become visible"

Looking east from Babylon on January 30, 999 B.C. at 6:34 A.M.

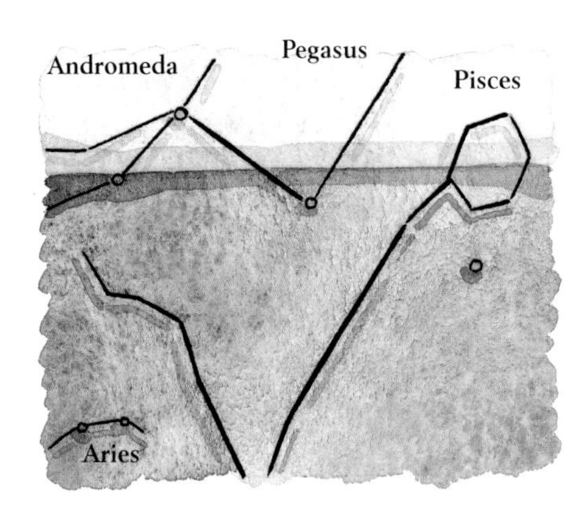

"On the 25th of Shabatu the Anunitu [eastern fish and part of the line of Pisces] becomes visible"

Looking east from Babylon on February 19, 999 B.C. at 6:14 A.M.

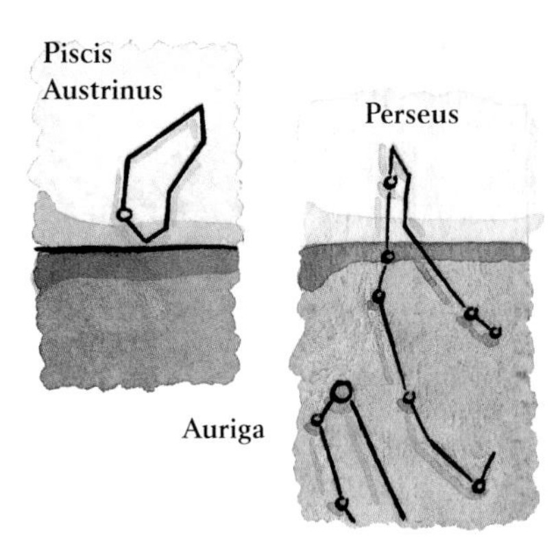

"On the 15th of Addaru the Fish [Piscis Austrinus] and the Old Man [Perseus] become visible"

Looking northeast (*far left*) and southeast (*right*) from Babylon on March 11, 999 B.C. at 5:57 A.M.

The origin of the zodiac

As a timekeeping device, the use of the rising and setting of stars was to have a long history, continuously from the Babylonian period down to the Middle Ages. In a sense, it has survived even to the present day in the use by astrology of the rising and setting of the zodiacal signs, although astrologers have kept the times of the astrological star signs artificially fixed since the Roman era. Because of a gradual shift of the earth on its axis, the dates for the signs of the zodiac are now out of synchrony with the constellations after which they are named. By about 500 B.C., the Babylonians saw the zodiac in its final form, a band across the sky divided into twelve segments (see diagram below). The fainter zodiacal constellations are hard to observe, but the Babylonians had an overriding aim in watching this particular belt of stars. These stars form a backdrop to the path of the sun (the "ecliptic"). The moon and the planets also move more or less along the ecliptic. The sun, the moon, and the planets moving in the foreground seemed to the Babylonians to be full of omen and portent. By association, the background of zodiacal stars also seemed to promise signals about what the future held.

The Babylonians divided the stars along the ecliptic in a second pattern as well–into twenty-eight equal segments called "mansions of the moon." The two brightest stars in the head of Aries are in the first mansion of the moon, the small stars in the belly of Aries are in the second, and so on. Astrological power was attached to the mansion in which the moon was rising.

Astrology among the Babylonians was a matter of searching the sky for warnings from the gods to the king (and by extension to society in general). Among the Greeks it became a system for telling individual fortunes on the basis (among other clues) of the individual's sun sign (which zodiacal constellation was rising on the morning of the day of the individual's birth) and star sign (which zodiacal constellation was rising at the moment of the individual's birth). The names of these constellations as they have come down to us (Aries, Taurus, Gemini, and so on) are simply Latin translations of earlier Greek names, which are in turn translations of Babylonian names.

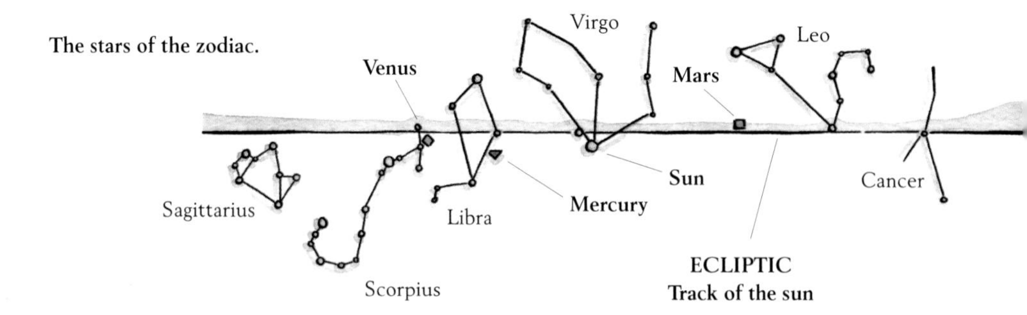

The stars of the zodiac.

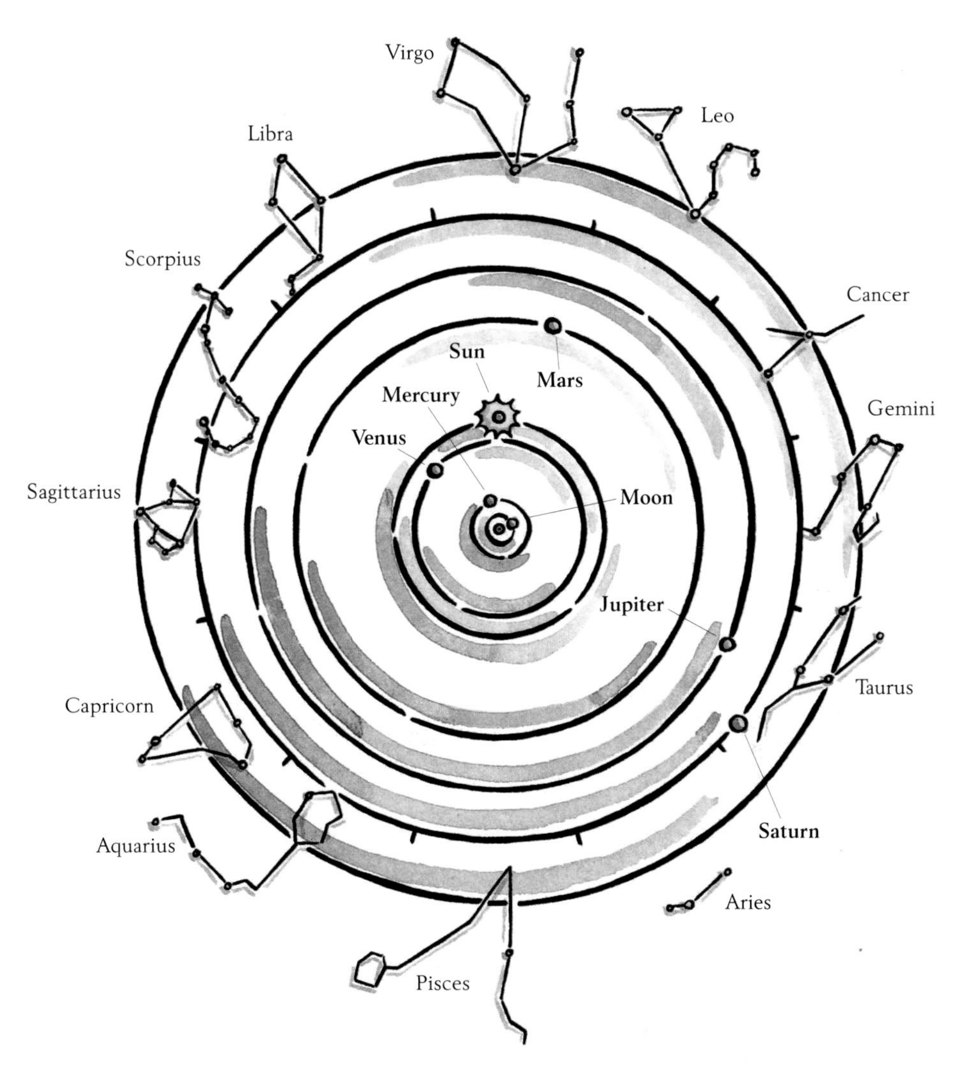

The zodiac is the path that the sun follows as it crosses the sky, the same path as the moon and the planets. The Babylonians divided this path into twelve. In each of the twelve segments they were able to discern a constellation, which they credited with astrological significance. They were the forerunners of the twelve signs of the zodiac used by astrologers today and of the celestial markers for zodiacal months in Greek calendars. These diagrams show the zodiac in linear form, as seen passing overhead at night, and set out in a medieval model of the universe with the earth at its center.

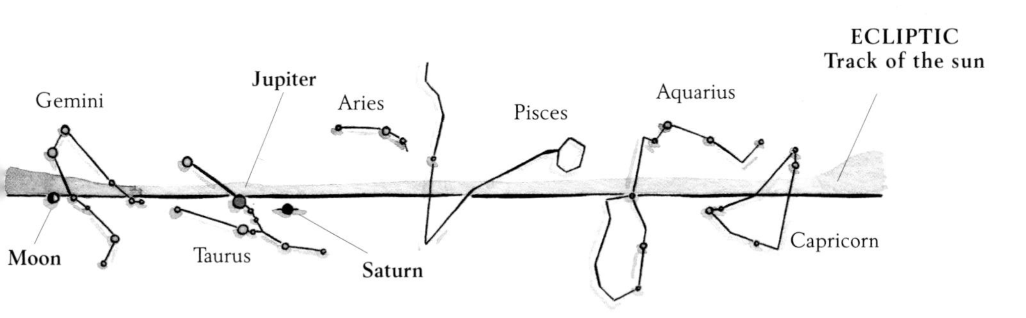

Next page: A zodiac from 1723 showing how the planets move between the houses of the zodiacal system.

11 Supremum Mobile Coelum
10 1. Coelum Chrystallinum
9 2. Coelum Chrystallinum
8 Firmamentum
7 Saturnus
6 Jupiter
5 Mars
4 Sol
3 Venus
2 Mercurius
1 Luna
Fig. I.
Terra
Cor Leonis
Leo
Orbita Satu
Spica Virginis
Virgo
30
Orbita Iovis
Libra
30
Scorpius
Fig. III.
30
Sagittarius
Aphelium
30
20
5000
100000
20000
30000
Semidiametri Terræ pro Planetarum distantijs.

Cancer
Gemini
Taurus 30
Aries
Pisces
Aquarius
Systema Ioviale
Fig II.
Fig. IV.
Saturnus
SOL
Iupiter
Mars
Terra
Luna
Mercurius
Venus

Greek star calendars

Our earliest records in Greek of the use of the stars as a chronological device occur between 750 and 700 B.C. In his agricultural poem, *The Works and Days*, Hesiod provides nine observations of the risings or settings of five stars or star groups. He also uses the summer and winter solstices as markers. In most cases, he times particular activities by reference purely to the occasion of the stars' rising, setting, or culmination in mid-sky. But in three instances he indicates the appropriate time for agricultural or nautical events by referring to the elapse of a certain number of days from a celestial event—for example:

> *Now, when Zeus has brought to completion sixty more winter days, after the sun has turned in his course, the star Arcturus, leaving behind the sacred stream of the ocean, first begins to rise and shine at the edges of evening. After him, the treble-crying swallow, Pandion's daughter, comes into the sight of men when spring's just at the beginning. Be there before her. Prune your vines.*

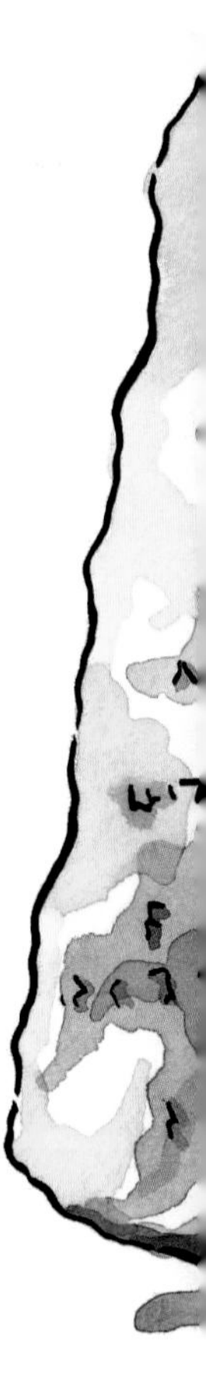

Hesiod's calendar is elegant in its efficient choice of very few celestial events to correspond directly with the pivotal periods of the agricultural year, and simple in that it demands no significant level of either literacy or numeracy.

After this relatively rudimentary, but seemingly effective, star calendar for farmers and sailors, the next stage is not encountered until the late fifth century B.C. in Athens. It was called a *parapegma*, and it was a formal means of keeping a record of the times of star-rise and star-set through the seasonal year. Its invention is connected with the names of two men, Meton (c. 440 B.C.–unknown B.C.) and Euktemon (fifth century B.C.), who are otherwise known for their work in astronomy.

The star-based *parapegma* in its developed form survives as inscribed stone tablets. Wood may also have been used but it has not lasted. Fragmentary examples have been discovered across the Mediterranean. They are organized in a number of columns on the stone slabs. A peg would be moved from one day to the next throughout the year, through a series of 365 holes. Alongside some holes are chiseled the observations of the stars for the day. Days lacking observations are marked simply by pegholes, set one after another in a line if necessary.

A stone *parapegma* from the ancient Greek city of Miletus in Asia Minor. The full, original calendar had 365 pegholes, some of which can be seen in this illustration. Each day of the year, the keeper of the calendar would move the peg on by one hole. Stellar observations inscribed on the calendar for certain days were a check against losing track and a check against seasonal drift. Probably every four years or so, the keeper would let the peg rest for a day (effectively creating a leap year) to allow stellar observations to catch up.

A peghole star calendar from Miletus

The following excerpt from a late *parapegma* from Miletus in western Turkey (about 100 B.C.) provides a typical example of this kind of calendar. Modern constellation equivalents are in square brackets.

• = peghole in the original *parapegma*

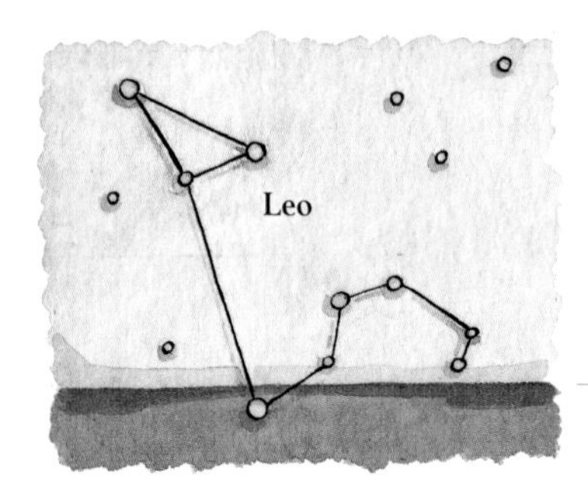

Looking west by northwest from Miletus on January 24, 100 B.C. at 7:28 A.M.

"• The Sun in the Water-Pourer [Aquarius]"

Horizon (below is not yet visible)

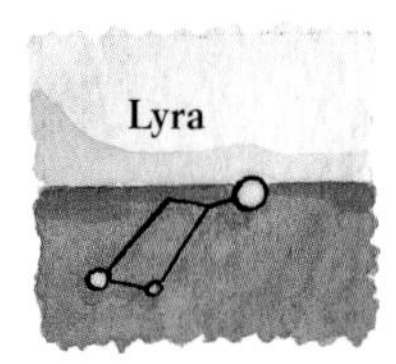

Looking northwest from Miletus on January 24, 100 B.C. at 6:07 P.M.

"• The Lion [Leo] begins setting in the morning and the Lyre [Lyra] sets • •"

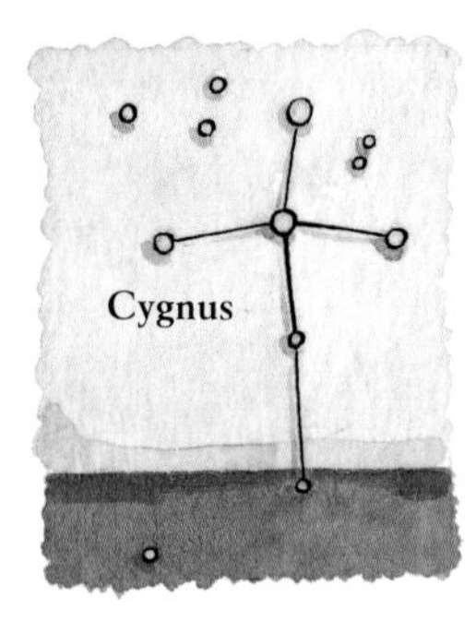

Looking west by northwest from Miletus on January 27, 100 B.C. at 5:29 P.M.

"• The Bird [Cygnus] begins setting at nightfall • • • • • • • • •"

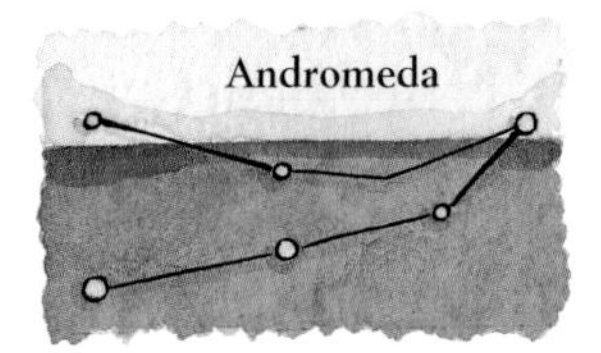

Looking east by northeast from Miletus on February 6, 100 B.C. at 6:46 A.M.

"• Andromeda begins to rise at dawn
• •"

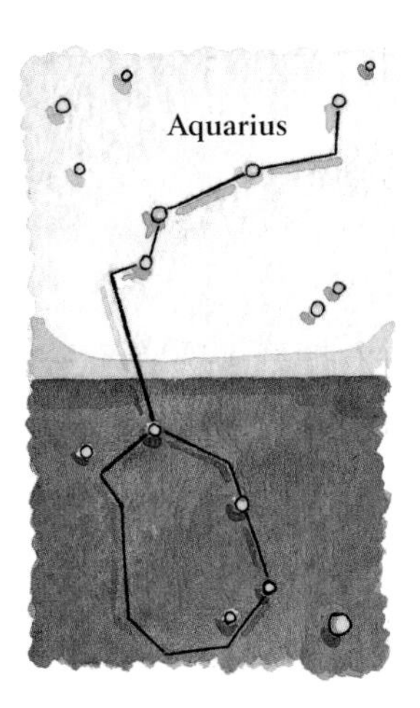

Looking east by southeast from Miletus on February 9, 100 B.C. at 6:46 A.M.

"• The Water-Pourer [Aquarius] is in the middle of its rising"

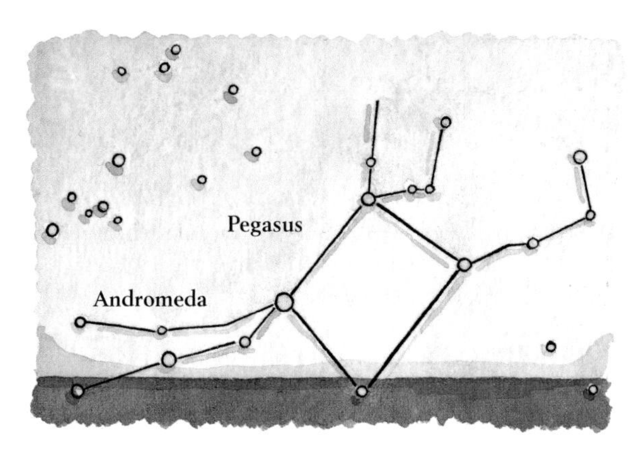

Looking east from Miletus on February 10, 100 B.C. at 7:09 A.M.

"• The Horse [Pegasus] begins to rise in the morning •"

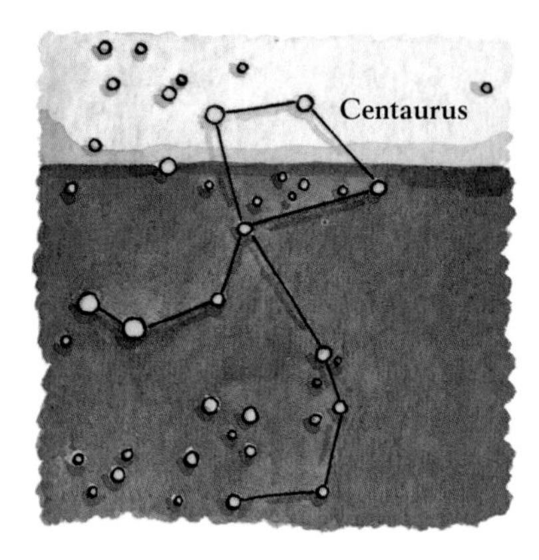

Looking southwest from Miletus on February 12, 100 B.C. at 7:09 A.M.

"• The whole Centaur [Centaurus] sets in the morning"

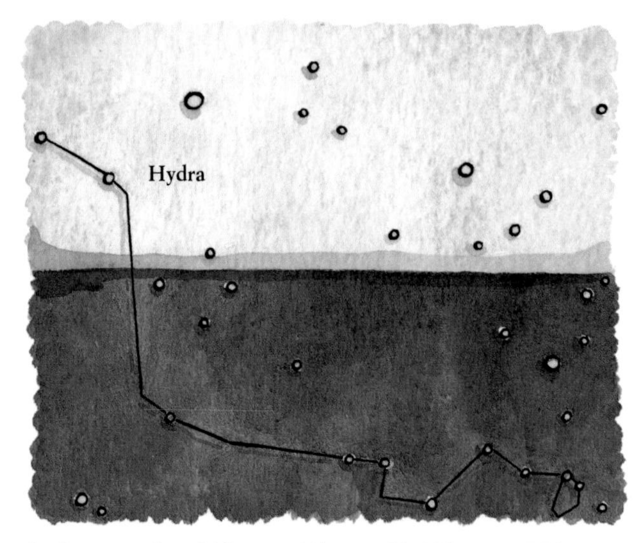

Looking west from Miletus on February 13, 100 B.C. at 7:06 A.M.

"• The whole Hydra [Hydra] sets in the morning"

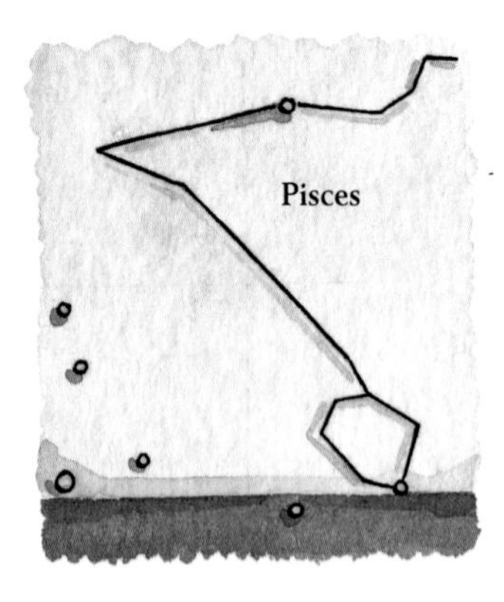

Looking west by southwest from Miletus on February 14, 100 B.C. at 5:47 P.M.

"• The Great Fish [Pisces] begins to set in the evening"

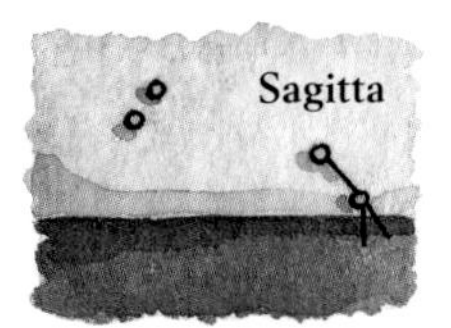

Looking west by northwest from Miletus on February 15, 100 B.C. at 3:36 P.M.

"• The Arrow [Sagitta] sets, a season of continuous west winds
• • • •"

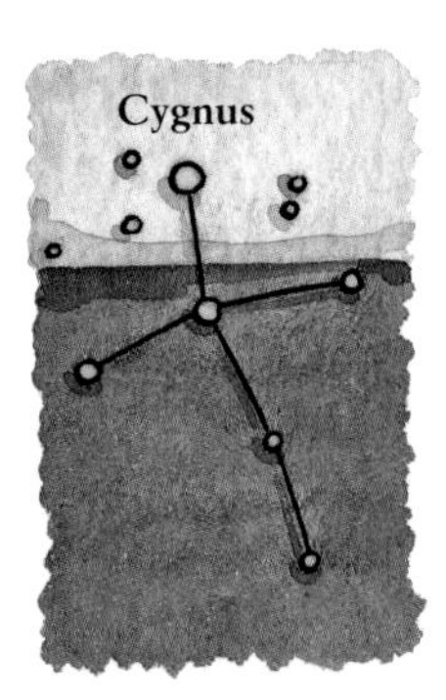

Looking northwest from Miletus on February 20, 100 B.C. at 5:53 P.M.

"• The whole Bird [Cygnus] sets in the evening"

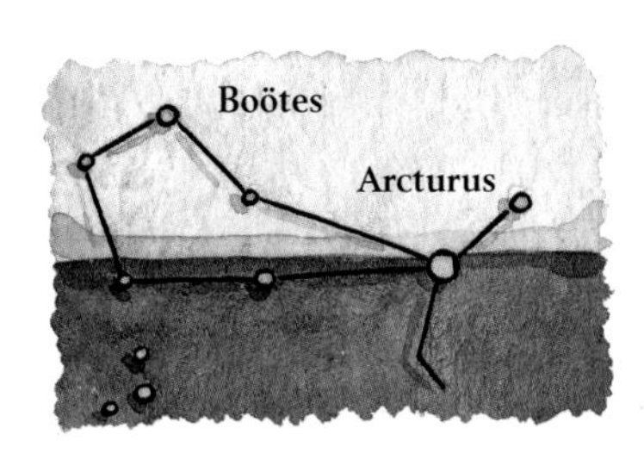

Looking northeast from Miletus on February 21, 100 B.C. at 7:02 P.M.

"•Arktouros [Arcturus] rises in the evening"

Zodiac signs painted on the inside of an Egyptian sarcophagus found at Thebes in 1823.

From star calendar to solar calendar

Of Meton's and Euktemon's original *parapegmas*, nothing survives, and evidence for Meton's in later writings is scant. But for Euktemon's there is a good deal of evidence from later writings, for it was one of the most popular star calendars then used. Most likely, Euktemon initially used only day counts between his sightings. However, at some later stage this *parapegma* was organized, probably by Euktemon himself, according to the zodiacal "months." Organizing observations of the stars into zodiacal months was a significant transition to a new type of calendar whose focus was more the sun's apparent annual path.

The Greeks were aware of the Babylonian division of the zodiacal belt into twelve thirty-degree segments by the later fifth century B.C., when they recognized the ecliptic (the path of the sun). It is possible that they derived the ecliptic from the Babylonian zodiac. By the early fourth century B.C., all twelve zodiacal constellations were recognized by the Greek astronomer Eudoxos (c. 408–c. 353 B.C.). He is supposed to have devised a type of sundial that recorded the passage of the sun into the zodiacal signs. Sundials like these had a long life in the Greek and Roman world, the most imposing, physically and politically, being perhaps the one erected by the Emperor Augustus in Rome in the late first century B.C.

Euktemon extended the traditional core stars that had been observed in Hesiod's time. Far more of his additions come from outside the zodiac than from within it, which suggests that a calendar with the sole scientific purpose of keeping track of the sun's apparent movement in the sky was not Euktemon's intention. For that, you would expect a list of stars more closely connected with the sun, like that in the *parapegma* of the Greek astronomer Kallippos a century later. His intention seems to have been practical timekeeping.

A calendar tied through the signs of the zodiac to the movement of the sun would be better able to keep pace with the seasons through the years, and would be a major step forward in the regulation of human affairs tied to the seasons. It is this type of calendar which some believe the Athenian historian Thucydides used in his history of the Peloponnesian War. It stands in stark contrast with most other types of calendar then in use in Athens and elsewhere in the Greek world, which, as lunar or lunisolar calendars, were easily put out of step with the seasons by haphazard intercalation by the civic authorities.

Meton published a calendar, based on a cycle of nineteen years, to make the lunar and solar years align in the way the Babylonians had done. This cyclical calendar was set up in cities beyond Athens by his followers, although there is no firm evidence from Athens that the cycle was used to regulate the lunisolar calendar there.

Egyptian calendars

When Greeks ruled Egypt (332–30 B.C.) they were in the process of developing calendars that owed much more to the Babylonians than to Egypt. The Babylonians' mathematics were more sophisticated than the Egyptians'. This gave precision to their astronomical observations, and made them the foundation Greeks chose to build on. However, for two thousand five hundred years before Greek rule, the Egyptians had been finding practical solutions to problems about keeping track of the days and years.

The earliest Egyptian calendars were based on the phases of the moon. The religious calendar divided the year into three seasons–the Nile flood (roughly July to October), planting and fresh growth (November to February), and the dry season (March to June), when the crops ripened and were harvested. Each season comprised four lunar months (some of twenty-nine, some of thirty days). But twelve lunar months are about eleven days short of the nearly 365.25 days in a seasonal year, so an extra month must have been added from time to time by decree when it seemed necessary to do so.

Egypt first became a single kingdom in about 3000 B.C. This political unification may have given the impetus to the development of a more regular system of timekeeping than could be provided by a lunar calendar. Alongside the religious calendar, an administrative calendar appeared soon after 3000 B.C. It efficiently avoided the difficulty that must have resulted in calculating the number of days between dates when some years were longer than others and had different numbers of months with different numbers of days. Each administrative year had exactly 365 days. There were twelve months, each of thirty days and named after a religious festival within it. Each month was neatly divided into three weeks of ten days each–so there were thirty-six weeks in the year. Five extra days, which didn't belong to any of the weeks, were needed to bring the total up to 365. These were what the later Greeks called the "epagomenal," or additional, days, and they were added at the end of the year.

Why 365 days were hit upon as the total for a year remains uncertain. Records suggest that the Egyptians were aware from an early stage that the seasonal year actually averaged about 365.25 days, but they did nothing to account for the extra quarter day every year, and so their administrative calendar drifted ever so slowly against the true seasonal year. Even much later, in 238 B.C., the decree of the Greek Egyptian king, Ptolemy III, to correct the wandering year by creating a leap-year system of adding an extra day every fourth year, was ignored by the priests who controlled the calendar. Such a system was imposed on Egypt in 26 B.C. by the Roman emperor Augustus, when he added a sixth epagomenal day to the Egyptian calendar every fourth year. It seems, then, that the fixed length of the Egyptian year had a symbolic significance which is lost to us now. Because no allowance had to be made for leap

Low relief zodiac from the Grand Temple at Denderah, c. 1826.

years in calculations, the Egyptian calendar was adopted for special purposes by the Greek astronomers who settled after 332 B.C. in the new Greek city of Alexandria in the Nile delta. It survived into medieval astronomy, and was used by Copernicus to construct tables of the motions of the moon and the planets.

But two natural events suggest ways in which the Egyptians could have arrived at a count of about 365 days for the year. The high point of the annual Nile flood could be measured by simple means–as it has been down to modern times by various Nilometers–to establish a gauge by day-counts of the length of the seasonal year. Such measurements could average out to the length of a seasonal year. The other event was the morning rising of the prominent star Sirius, called Sothis by the Egyptians. They noticed that the first observed morning rising (the heliacal rising) of Sothis, after a period of seventy days of invisibility, coincided with the start of the Nile flood. The day of this event became the the first day of the month of Thoth, the start of the Egyptian year. Over a long period of time, it could be seen that the average time between the heliacal risings of Sirius was 365.25 days, practically the length of the seasonal year. However, while the rising of Sothis may have formed the basis of the administrative calendar at its inception, once that calendar was established, the actual day of the rising ceased to be a marker for the start of the new year, because the calendar held to just 365 days.

It seems that by about 2000 B.C. the Egyptians had also picked on the heliacal risings of thirty-five other stars or star groups as especially significant along with that of Sirius. A given heliacal rising was used to herald the coming dawn. After ten days, a given star would have lost its usefulness, owing to the slippage between the solar and the stellar years (the sun appears to slip away from the star about four minutes every night), and would be replaced by the next star in the list. These thirty-six stars are now known as the decans, after the Greek name for them (*deka* is Greek for ten). Records of the decans are so poor that, apart from Sirius, it is not possible to identify which stars they were.

Drift in the old Roman calendar

In spite of adding an extra month (Mercedinus) every other year, there was still a drift in the Roman calendar.

Roman dates	Gregorian dates
1–29 Ianuarius	January 1–29
1–28 Februarius	January 30–February 26
1–31 Martius	February 27–March 29
1–29 Aprilis	March 30–April 27
1–31 Maius	April 28–May 28
1–29 Iunius	May 29–June 26
1–31 Quintilis	June 27–July 27
1–29 Sextilis	July 28–August 25
1–29 September	August 26–September 23
1–31 October	September 24–October 24
1–29 November	October 25–November 22
1–29 December	November 23–December 21
1–29 Ianuarius	December 22–January 19
1–23 Februarius	January 20–February 11
1–27 Mercedinus	February 12–March 10
1–31 Martius	March 11–April 10
1–29 Aprilis	April 11–May 9
1–31 Maius	May 10–June 9
1–29 Iunius	June 10–July 8
1–31 Quintilis	July 9–August 8
1–29 Sextilis	August 9–September 6
1–29 September	September 7–October 5
1–31 October	October 6–November 5
1–29 November	November 6–December 4
1–29 December	December 5–January 2
1–29 Ianuarius	January 3–January 31
1–28 Februarius	February 1–28
1–31 Martius	March 1–31
1–29 Aprilis	April 1–29
1–31 Maius	April 30–May 30
1–29 Iunius	May 31–June 38
1–31 Quintilis	June 29–July 39
1–29 Sextilis	July 30–August 37
1–29 September	August 28–September 25
1–31 October	September 26–October 26
1–29 November	October 27–November 24
1–29 December	November 25–December 23
1–29 Ianuarius	December 24–January 21
1–23 Februarius	January 22–February 13
1–27 Mercedinus	February 14–March 12
1–31 Martius	March 13–April 12
1–29 Aprilis	April 13–May 11

The Roman year began on March 1. The months were divided into three parts—at day one (*kalendae*), and depending on the actual month, at day five or seven (*nonae*), and day thirteen or fifteen (*idus*). These divisions represent notional lunar phases from new moon (the name *kalendae* derives from the proclaiming of the new crescent by the priest), to first quarter (the name *nonae* simply signifies eight days—nine by Roman inclusive reckoning—before the next division), to full moon (the name *idus* may stem from a Greek word for the full moon). The remaining days of the month were numbered according to their relationship to one of these three divisions, using inclusive and retrospective reckoning. For example, January 2 was designated *ante diem IV non. Jan.*—the fourth day before the *Nones* of January (with the *Nones* falling on January 5).

There are signs in the names of the Roman months that the calendar was originally made up of ten months—the names September through December representing literally the seventh through tenth months. By the first century B.C., the calendar comprised twelve months, four of thirty-one days (*Martius*, *Maius*, *Quintilis*, *October*), seven of twenty-nine days (*Ianuarius*, *Aprilis*, *Iunius*, *Sextilis*, *September*, *November*, *December*), and one of twenty-eight days (*Februarius*), a total of 355 days, with an intercalary month of twenty-two or twenty-three days (called *Mercedinus* or *Mercedonius*) inserted after February 23 every second year to keep with the solar year. The five remaining days of February were added to this month, to give it twenty-seven or twenty-eight days.

In 44 B.C., before Julius Caesar's assassination, the Roman senate decreed that the month *Quintilis* should be called *Iulius* after him, because he was born in that month. The month *Sextilis* was named *Augustus*, in the lifetime of the emperor of that name, because it was the month in which he gained his most significant political honors. Later emperors tried to change the names of the months—most notoriously Commodus, who wanted to change the name of every month after some aspect of himself—but these alterations did not take hold. The sequence of the Roman months was the same as it still is in the Western calendar, allowing for anglicizations: January, February, March, April, May, June, July, August, September, October, November, December.

Adding an extra month every two years made a four-year cycle of Roman years four days longer than four solar years, and the calendar grew badly out of step with the solar year. This type of intercalation was abandoned at some unknown time, and instead the Romans, like the Greeks, practiced intercalation on a haphazard basis. Attempts were made from the mid-fifth century B.C. to get back into phase with the seasons, but to no avail. Over the last two centuries B.C., the divergence of the calendar from the seasonal year stretched from as many as 117 days in 190 B.C. to near correspondence between 140 and 70 B.C. By the time of Julius Caesar, the discrepancy had grown again to ninety days.

The Julian calendar

In 46 B.C. Caesar ordered a wholesale revision of the calendar to the point of adopting a purely solar one covering 365.25 days, the quarter-day to be absorbed into an extra single whole day added every fourth "leap" year. This solar year had been discovered earlier, but had never been put to use.

At about 330 B.C. the Greek astronomer Kallippos (c. 370–310 B.C.) discovered that the year was made up of about 365.25 days. He proposed combining four Metonic cycles of nineteen years each into a period of seventy-six years from which one day was to be dropped. This gave a total of 27,759 days over seventy-six years, and therefore an average year of 365.25 days. Kallippos's formula relied on intercalated months, rather than single days, being added over time.

In 238 B.C. the Greek-Egyptian King Ptolemy III Euergetes had decreed the addition of an extra day every fourth year into the Egyptian calendar to correct the drift from the solar year, but this did not take root. Significantly, in his own calendar reform Caesar had the services of the Greek astronomer Sosigenes, who came from Alexandria in Egypt.

It was Julius Caesar who decreed that the calendar be entirely revised.

Ninety days were added by Caesar to 46 B.C., making it 445 days long. This was achieved by inserting not only the intercalary month after February but two further such months between November and December. He called this year the "final year of confusion." From January 1, 45 B.C., a common year of 365 days was instituted, with months of the same length as they are nowadays in the Western calendar; the extra ten days over the former 355-day year being placed at the end of different months so as to maintain the usual dates for religious festivals. Thus a festival formerly held on December 21 remained on that day, although its notation changed from *X kal. Jan. to XII kal. Jan.* because the month now had thirty-one days instead of thirty. Every fourth year an extra day was meant to be inserted after *VI kal. Mart.* (that is, February 24), the new day being called *bis sextum kal. Mart.*; hence the "leap" year was called *annus bissextus*, from which comes the English term "bissextile" for a "leap" year. After Caesar's assassination in 44 B.C., the priests initially inserted the extra day by mistake every three years, leading Augustus in 9 B.C. to omit any further intercalation for sixteen years. The Julian year began to function properly from only A.D. 8.

46 B.C.– the final year of confusion

Month	Number of days
Ianuarius	29
Februarius	23
Mercedinus	28
Martius	31
Aprilis	29
Maius	31
Iunius	29
Quintilis	31
Sextilis	29
September	29
October	31
November	29
Intercalary	33
Intercalary	34
December	29
	445

45 B.C.– the first year of the Julian calendar

Month	Number of days
Ianuarius	31
Februarius	28
Martius	31
Aprilis	30
Maius	31
Iunius	30
Quintilis	31
Sextilis	31
September	30
October	31
November	30
December	31
	365

The seven-day week

A further innovation in the Roman calendar which has lasted through to the present day was the introduction of the seven-day week. This was a late feature, introduced in the Emperor Augustus's time, replacing an original eight-day, market-oriented week, with seven days of work and the eighth for the market. The seven-day week as the Western world knows it (Saturday, Sunday, Monday, Tuesday, and so on) reflects a variety of influences, including Mesopotamian and Greek planetary astronomy, and the Judaic concept of the Sabbath.

The weekday names are derived from those of the seven planets: Saturn, Jupiter, Mars, Sun, Venus, Mercury, and the Moon. This sequence corresponded to their distance from the earth, and in turn it governed the sequence of the hours of the day over which each planet was said to rule. Thus, Saturn ruled the first hour of Saturday, Jupiter its second, Mars its third, and so on to the twenty-fourth hour, which Mars again ruled. The first hour of the next day was then ruled by the Sun (hence, Sunday), the second by Venus, and so on to the twenty-fourth, which was ruled by Mercury. The first hour of the next day was therefore ruled by the Moon (hence, Monday), and so on. The eventual sequence of the planets as rulers of successive days of the week was therefore Saturn, Sun, Moon, Mars, Mercury, Jupiter, and Venus.

In English, but more so in Romance languages, some of these names are still reflected in the names of the weekdays—for example, in French, *mardi* (Tuesday) from Mars, and *mercredi* (Wednesday) from Mercury; but in English the equivalent Norse gods' names have superseded most of the original Roman ones, to give us our present names, Tuesday from Tiw, Wednesday from Wodin, Thursday from Thor, and Friday from Frig.

Easter tables

The outstanding calendrical question of the medieval world was the calculation of the date of Easter, the crucial event of Christianity. Easter Sunday is usually defined as the first Sunday after the first full moon following the spring equinox. This definition goes back to Christ's death and resurrection at the time of the Jewish festival of Passover, as reported in the gospels. Passover is itself related to the date of the spring equinox and the middle of the Jewish month of Nisan, and therefore to a full moon. However, neither "full moon" nor "spring equinox" are what they normally are in an astronomical context; both have ecclesiastical definitions which separate them from anything observational. The ecclesiastical "full moon" is a calculated and artificial phenomenon, based on various more or less inaccurate cycles, such as the nineteen-year Metonic cycle, rather than on observation. The ecclesiastical "spring equinox" is fixed as March 21, regardless of the fact that the phenomenon shifts up and down the calendar by two or three days. This date was fixed by the first Christian Ecumenical Council, the Council of Nicaea (modern Iznik in northeastern Turkey) in A.D. 325, to put an end to controversies in the Church over this crucial date in its calendar.

A pivotal figure in the definition of Easter, not often remembered nowadays, was the monk Dionysius Exiguus (c. sixth century A.D.). In A.D. 525 he prepared a set of Easter tables which built on those of a number of predecessors. In the mid-fifth century A.D. Cyril, patriarch of Alexandria, had established a ninety-five-year Easter table, based on five nineteen-year Metonic cycles. Dionysius took the last cycle (A.D. 513–31) from Cyril's table, and added a ninety-five-year cycle to it, thus extending the

The Roman week and the gods

Each hour is ruled by a god. Each day takes its name from the ruler of the first hour.

	SAT	SUN	MON	TUES	WED	THU	FRI
1	Saturn	Sun	Moon	Mars	Mercury	Jupiter	Venus
2	Jupiter	Venus	Saturn	Sun	Moon	Mars	Mercury
3	Mars	Mercury	Jupiter	Venus	Saturn	Sun	Moon
4	Sun	Moon	Mars	Mercury	Jupiter	Venus	Saturn
5	Venus	Saturn	Sun	Moon	Mars	Mercury	Jupiter
6	Mercury	Jupiter	Venus	Saturn	Sun	Moon	Mars
7	Moon	Mars	Mercury	Jupiter	Venus	Saturn	Sun
8	Saturn	Sun	Moon	Mars	Mercury	Jupiter	Venus
9	Jupiter	Venus	Saturn	Sun	Moon	Mars	Mercury
10	Mars	Mercury	Jupiter	Venus	Saturn	Sun	Moon
11	Sun	Moon	Mars	Mercury	Jupiter	Venus	Saturn
12	Venus	Saturn	Sun	Moon	Mars	Mercury	Jupiter
13	Mercury	Jupiter	Venus	Saturn	Sun	Moon	Mars
14	Moon	Mars	Mercury	Jupiter	Venus	Saturn	Sun
15	Saturn	Sun	Moon	Mars	Mercury	Jupiter	Venus
16	Jupiter	Venus	Saturn	Sun	Moon	Mars	Mercury
17	Mars	Mercury	Jupiter	Venus	Saturn	Sun	Moon
18	Sun	Moon	Mars	Mercury	Jupiter	Venus	Saturn
19	Venus	Saturn	Sun	Moon	Mars	Mercury	Jupiter
20	Mercury	Jupiter	Venus	Saturn	Sun	Moon	Mars
21	Moon	Mars	Mercury	Jupiter	Venus	Saturn	Sun
22	Saturn	Sun	Moon	Mars	Mercury	Jupiter	Venus
23	Jupiter	Venus	Saturn	Sun	Moon	Mars	Mercury
24	Mars	Mercury	Jupiter	Venus	Saturn	Sun	Moon

table to A.D. 626. The Dionysiac tables were later extended further to A.D. 721. Dionysius Exiguus also changed the fundamental epoch from which the years since Christ had been calculated. The Church in Alexandria had used as its epoch the date when Diocletian became emperor of Rome (in our terms A.D. 284). But this emperor was notorious in the Church for a severe persecution that he instigated against the Christians. To remove the connection between the ecclesiastical calendar and one of the Church's worst enemies, Dionysius sought another epoch, calculating backwards in time to the incarnation (conception) and nativity (birth) of Christ.

It is not entirely clear how he did this. In the Gospel of Luke (3: 1–23) the start of Christ's ministry is dated to about the time he was thirty years old, in the fifteenth year of the reign of the Roman Emperor Tiberius. In Roman terms, this year was 782 A.U.C. The initials A.U.C. indicate the traditional Roman epoch; they stand for the Latin *ab urbe condita*, which means "from the foundation of the city," and refers to the supposed foundation date of Rome, which took place in 753 B.C. in our terms. Apparently working back from 782 A.U.C., Dionysius subtracted twenty-nine full years from this date, and arrived at 753 A.U.C. for the year of Christ's incarnation on the traditional Christian date of March 25 and his nativity on December 25.

By coincidence, this year of 753 A.U.C. was also the date arrived at by subtracting 532 years from the time of Dionysius. This is a significant figure in the context of establishing the date of Easter, a task close to the heart of Dionysius. The "great year" of 532 years had been devised about seventy years before Dionysius by Victorius, Bishop of Aquitaine in western France. It provided a means of coordinating the lunar and solar calendars together with the seven-day week cycle, to bring not only the same phase of the moon back to the same date in the solar calendar but back to the same weekday as well. If Easter was tied solely to a full moon and therefore to the lunar calendar, it would run, just as Islamic holidays still do, through all the seasons. But Easter is tied both to the full moon and to the equinox, a solar event. To coordinate the lunar and solar calendars, the nineteen-year Metonic cycle was devised. Easter, however, is also tied to a particular day of the seven-day week, Sunday. So Easter tables had to show a way of combining the lunar, solar, and weekday cycles. The 532-year cycle provided the means. Combining the weekday cycle of seven days with a four-year cycle of the Julian calendar ($7 \times 4 = 28$) acknowledged that every twenty-eighth year a given solar calendar date would recur on the same weekday—for example, if January 1 fell on a Tuesday in A.D. 513, it would do so again in 541; the use of the four-year cycle in this calculation allowed for the additional leap day every fourth year. Combining this twenty-eight-year cycle with the nineteen-year Metonic cycle ($28 \times 19 = 532$) brought a desired phase of the moon into line with a solar date on a given weekday.

This coincidence between biblical evidence and calendrical calculation for Easter may also have led Dionysius to believe that 753 A.U.C. was indeed the year of Christ's birth, and therefore the appropriate starting date for the Christian era. While the year of Christ's birth is still a matter of debate, it is generally accepted that the scraps of historical evidence available to us tend towards a date more in the region of 4 B.C. than 1 B.C.

The year 753 A.U.C. corresponds to our 1 B.C., and the following year to our A.D. 1. However, the regular use of "A.D." (*anno Domini*, Latin for "in the year of the Lord") did not begin until the eighth-century English monk and historian, the Venerable Bede, adopted it; "B.C." made its appearance much later, in the seventeenth century.

nglo-Saxon scholar ne Venerable Bede ntroduced the use of ne term *anno Domini*.

The drifting Julian calendar

The Julian year initiated in 45 B.C. has one significant fault: at an average of 365.25 days, it is still slightly too long to represent the true astronomical year, and, as a result, to ensure that the astronomical reference points of the solstices and equinoxes stay on more or less the same dates every year. In terms of the predominant Christian calendar from the Middle Ages onwards, this proved an important issue because of the connections between the date of Easter and that of the spring equinox, and between the solstices and the birthdays of John the Baptist and Christ (June 25 and December 25, according to scriptural commentaries). Also, since the full moon is a central feature of the calculation of Easter, the inaccuracy of the nineteen-year Metonic cycle for the moon's phases became apparent and attracted attention over time. Over a sufficiently long period of time, the Julian calendar allowed the religious and seasonal years to drift apart.

Attempts to address this problem were made from the Middle Ages onwards. The acquisition by Western Europe of both Arabic and ancient Greek astronomical texts alerted scholars like John of Sacrobosco and Roger Bacon in the thirteenth century to the discrepancies in the current calendar. Bacon wrote to Pope Clement IV on the issue. In the late fifteenth century the German scholar Regiomontanus (Johann Müller of Königsberg) was invited to Rome to assist in the reconstruction of the calendar by Pope Sixtus IV, but he died before much was done. Interest continued at the highest levels in the Church through the early sixteenth century, but to no end (one proposal at this time recommended allowing the spring equinox to shift through the calendar, rather than attempting to fix its date).

The Gregorian calendar

The Council of Trent (1545–63) finally authorized the pope to instigate calendar reform. Nine years later, in 1572, Ugo Boncompagni, deputy to Pope Pius IV at the Council of Trent, became pope, taking the name of Gregory XIII. It was in his reign that the reform finally took place. The calendar commission was convened in the mid-1570s. Aloysius Lilius, a physician-astronomer, devised the elegantly simple solutions which were eventually adopted by the commission.

Lilius argued for the length of the year being based on the motion not of the real sun but of an imaginary mean sun. This mean sun is the average position which the sun would occupy if the earth's orbit was circular (and not elliptical) and if the earth was perpendicular to the plane of its orbit (instead of tilting at 23.4 degrees to it). Days of the mean sun are of constant twenty-four-hour length, whereas real days of the real sun vary over a year. So the use of the mean sun simplified matters for Lilius (as it

The drifting spring equinox in the Julian calendar

Year	Julian date for the spring equinox
300	March 20
700	March 17
1100	March 14
1500	March 11

still does for modern astronomers). The process of simplification went further—Lilius did not remeasure the sun's apparent motion for a value of the mean, but instead derived it from a set of popular astronomical tables, the Alfonsine Tables, written originally in 1252 for King Alfonso X of Castile in Spain and updated over the years. The value of the year thus derived was 365 days, five hours, forty-nine minutes, sixteen seconds—the equivalent of 365.2425 days, and only thirty seconds longer than the true year. This Alfonsine year runs short of the Julian by ten minutes forty-four seconds, which is equal to one day every 134 years. To compensate for this in a reform which maintained the Julian year in all other respects, three days every 402 years (134 years x 3) would need to be dropped, or (to keep things simple) three days in 400 years. These three days were nominated as the leap years in three out of four century years (that is, those not divisible by four). This reduction gives an average Gregorian year of 365 days, five hours, forty-nine minutes, twenty seconds.

More work on the date of Easter

Since the Julian reform, the calendar had drifted from the true year by several days. If ten days were recouped in the reform, then the spring equinox would not fall later than March 21, the date assigned to it by the Council of Nicaea in A.D. 325, thus bringing the fixed ecclesiastical and variable astronomical equinoxes reasonably close together again. At the time of the calendar reform, this was more important than getting the solar year correct, because the impetus for the reform was fueled more by a desire to calculate the date of Easter correctly. With the date of Easter being based on the full moon following the spring equinox, the Church wanted to know in advance that the ecclesiastical and astronomical moons and equinoxes would behave according to its need, with the full moon (ecclesiastical and astronomical) following the equinox,

***Next page*: A discussion of the reform of the calendar, which took place under Pope Gregory XIII.**

and not vice versa. The calculation of the moon's phases had long been based on the assumption that the nineteen-year Metonic cycle was accurate: the sun and moon, it was understood, were brought back into the same positions that they held nineteen years earlier, and would do so again every nineteen years hence, because nineteen solar years equaled a complete number (235) of lunar months. In fact they do not, but run ahead of the moon's cycles by about an hour-and-a-half in that time (nineteen mean Julian years make 6,939 days, eighteen hours, while 235 mean lunations make 6,939 days, sixteen hours, thirty-one minutes). Arab and eventually European scholars had noted this since the Middle Ages. Lilius calculated that the moon was drifting against the Church's lunisolar calendar by one whole day every 312.7 years, and, by the 1570s, four whole days' difference lay between the ecclesiastical moon and the astronomical. He worked out that eight periods of 312.7 years equals just over 2,500 years, a number which can be divided into seven periods of 300 years plus one period of 400 years. He recommended dropping one day from the lunar calendar every 300 years seven times, and then an additional eighth day after 400 years, all to be done at the end of appropriate centuries.

These calculations allow Easter to wander through a determinable period of weeks over the years as the time-lag between full moon and equinox varies, and people naturally talk, for instance, of an "early" or "late" Easter. By contrast, other Christian festivals are tied to the solar calendar—most obviously, Christmas, whose date, December 25, marked in the ancient Roman calendar the birthday of the Sun at the winter solstice.

A slow reform

The Gregorian reform of the Julian calendar was driven by religious concerns, and particularly by the perceived need to regularize the timing of Easter within the context of a lunisolar calendar. Because of its religious character, the reform also had a political dimension in a Europe split along religious lines between Catholic and Protestant countries.

The reform's adoption across Europe and in the new colonies in the Americas was therefore a slow process. The decree to bring about the changes recommended by the calendar commission was signed in February 1582, with its adoption expected by October that year. Italy, Spain, and Portugal met this expectation. Other Catholic countries—France, Belgium, and the Catholic states of the Netherlands—delayed until December that year (so much so that Belgium missed Christmas Day that year), while the German states in the Holy Roman Empire converted to the new calendar over the period 1583–84.

Protestant countries rejected the reform, a situation that made travel across the German states particularly confusing, with the "Old Style" calendar preserved in the Protestant towns and the "New Style" adopted in the Catholic. It was not until 1700 that Protestant Germany and Denmark finally adopted the reformed calendar, although they calculated Easter differently until 1775. England (with her colonies) delayed until 1752. Sweden adopted the reforms piecemeal between 1700 and 1753.

The Eastern Orthodox churches rejected the reform until after 1923. Most are now aligned with the new calendar, but even so they retain their traditional method of calculating Easter, so that it is still possible to experience two Easters in the same year in Europe. In Asia, Japan adopted the calendar in 1873 during its westernization period, while China held out until 1912, although it was not until the Communist victory in 1949 that the whole country was made to adopt it.

5: exotic timekeepers

Clive Ruggles

Alternative calendars

*Ezeulu's announcement that…
the six villages would be locked in the
old year for two moons longer spread
such alarm as had not been known…
in living memory* Chinua Achebe

Did all early timekeeping develop in the same way, with different human societies progressing to different stages along much the same path as led, in Chapter Four, to our own calendar? The sequence of development would be that a society begins with a lunar calendar, then introduces extra months to keep in line with the seasons. This would evolve into using the risings and settings of stars to increase seasonal precision, and finally the society would work out a solar mean to govern a calendar whose months are completely detached from phases of the moon. But is this sequence inevitable?

Different ways of marking time

A wealth of human societies around the world, both in the past and in the present, show other paths of calendar development. The Barasana are a group of hunter-gatherers living in the Colombian Amazon. Their view of the world is very different from ours. To them, the sacred He world exists alongside the everyday world, and its chief protector, "Old Star," has many different manifestations. He is all at once a human warrior, the constellation Orion, the fierce thunder jaguar, and a short trumpet. What we would see as myth and reality are, to the Barasana, inextricably mixed.

At a certain time of year, food sources are scarce and a vital one comes in the form of caterpillars that fall down from a particular type of tree. The Barasana know when to move to the appropriate area of the forest, at least in part because they observe the "Caterpillar Jaguar" constellation, which corresponds to our constellation of Scorpius. The time of year when Scorpius rises higher and higher in the eastern sky after dusk coincides with the time of year when many species of Amazonian butterfly and moth breed; and as the caterpillars pupate they come down from the trees on which they feed.

This explanation comes from a Westernized understanding of what is happening. The Barasana explanation would be entirely different. They would say that the Caterpillar-Jaguar constellation is the father of caterpillars, and is responsible for the increase in their numbers as he rises higher and higher in the sky at dusk. The Barasana perceive a direct connection between the celestial caterpillar and terrestrial caterpillars, and this is reinforced by what is observed year after year. Their understanding of the world may be entirely different from ours, but it clearly results in a reliable method of timekeeping, and one which is a vital aid to survival at a difficult time of year. This is a notable achievement in its own terms.

This may seem to be far removed from the latest methods of accurate timekeeping employed by modern science. But it is relevant to everyone. We should not just aim to understand (in our terms) how a given method

Antares, the giant red star that is the brightest in Scorpius.

People who ignore the sun

Not 125 miles from the Mursi lives a community with a completely different, and very unusual, method of timekeeping. The Borana are nomadic cattle-herders whose territory straddles the border between Ethiopia and Kenya. In direct contrast to the Mursi, the Borana schedule their lives to a considerable degree. Rituals, ceremonies, and political and economic activities are all closely regulated by the calendar.

This calendar was studied in the 1960s by an Ethiopian anthropologist, Asmarom Legesse, who consulted with Borana experts on sky observation, known as *ayantu*. Some aspects of his account are not surprising. The Borana calendar is based upon lunar cycles, and there are twelve named months, the months being marked by the appearance of the new crescent moon. Furthermore, the calendar is carefully kept in check (at least for half of the year) by observing the position of the new moon relative to one of seven stars or star groups—Triangulum, the Pleiades, Aldebaran, Bellatrix, Orion's belt and sword, Saiph, and Sirius. During the other half of the year the *ayantu* identify the moon at different phases in relation to a particular star group, Triangulum. This in itself is interesting, since we tend to assume that any human community is likely to have attached most significance to the brightest stars; yet here is one for whom the relatively dim stars of what for us is a very minor constellation provide a pivotal reference point for the calendar.

The most curious thing about Borana timekeeping is a cycle of twenty-seven named days that runs independently of the sequence of lunar months. For many years this puzzled interpreters: why have a day-name cycle with a period so close to that of the monthly cycle? Each month lasts twenty-nine or thirty days, and the starting day of each month is two or three days later than that of its predecessor. In any given month certain days appear twice, once at the beginning and once at the end.

Part of the answer becomes clearer when we realize that the Borana live five degrees from the equator, so that the characteristic motion of the heavenly bodies is close to vertical: the sun, moon, and stars appear to rise straight upward in the eastern sky and fall straight downward in the west. There are two other vital clues. Several of the months have names that are similar to the Borana names of stars or star groups. Second, twenty-seven days is close to the period of the sidereal month—the time it takes the moon to move around the sky with respect to the stars.

The solution is that Borana experts keep track of the day by identifying stars that rise or set on a level with the moon on any night, regardless of its phase. (Only on nights very close to a new moon is there no possibility of seeing the moon at any time in the night.) They have twenty-seven "reference" stars or star groups, and the day is determined

a Navajo tribal fair:
and painting of the
arters of the world.

by identifying which one is on a level with the moon on any given night. The month is determined by observing which star or star group the new crescent moon sets on the level with. In this way, the *ayantu* keep track of the day and month from the moon and stars, without reference to the sun.

Quartering the world

For us, space and time are separate entities. But the distinction is blurred for many indigenous peoples, and keeping the calendar is intricately related to organizing the landscape. The inhabitants of the traditional Hopi village of Walpi, in Arizona, continue this tradition to the present day. From here, the rising position of the sun is carefully tracked against distant landmarks in order to ensure the correct timing of a variety of ceremonies which are seen to keep nature's equilibrium and ensure rain, good harvests, good health, and peace.

The elaborate nature and remarkable precision of Hopi horizon calendars fascinated visitors from the nineteenth century onwards. Yet Hopi villagers had more than just a sacred calendar related to crop-planting activities that was regulated by horizon sun observations. Walpi, in common with other villages, was seen as the very center of the world. The points on the horizon behind which the sun rises and sets at the solstices are themselves sacred places, which are still visited at certain times of the year for prayer sticks to be offered to the sun. They are important because they mark the four directions that are sacred in the Hopi conception of the world. This idea of our planet being divided into four parts, marked out by either the directions of both solstice days or the points of the compass, is widespread both amongst indigenous communities in North America, and was also common amongst the ancient civilizations of Mesoamerica.

The directions of the solstice days are not geometrical abstractions to the Hopi, but fundamental axes that help define sacred space; empirical realities with rich symbolic associations. Observations from Walpi of the sun rising and setting at the solstices are actually superfluous to the

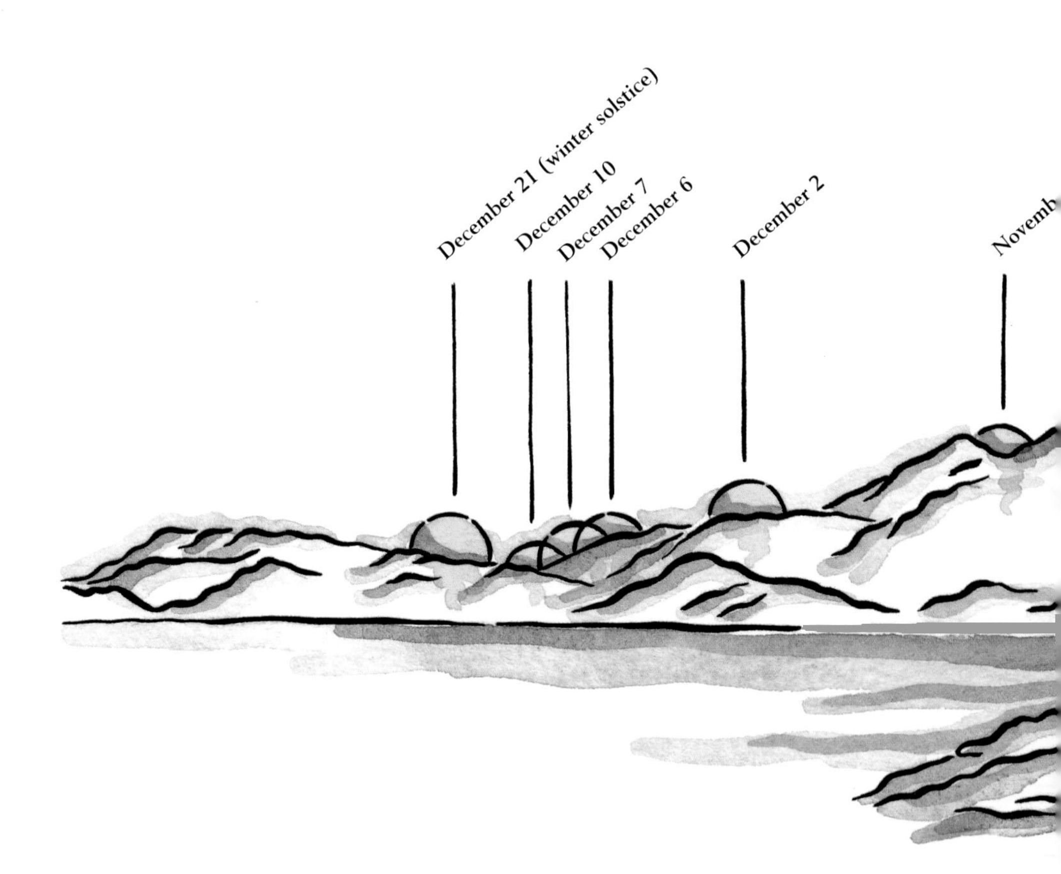

Moving with the sun

Many indigenous societies perceive there to be an intimate, spiritual connection between people and their natural environment, both the land and the sky. Places in the landscape, such as prominent landmarks, acquire particular significance and meaning, and the power and vitality of being in a particular place–perhaps at a particular time–is reinforced by myth and oral tradition, as well as ritual practices that take place there.

Sometimes myth and ritual regulate seasonal movements through the landscape that are necessary for survival. The traditional subsistence cycle of the Lakota people of South Dakota involved following buffalo herds around the famous Black Hills. The Lakota directly associated landmarks on the ground with particular groups of stars in the sky. The buffalo was seen as the embodiment of the power of the sun, and they ensured that they were in the right place at the right time by keeping their own progress through the hills in time with that of the sun through the relevant constellations. As the sun moved through the Lakota constellations it gave the people spiritual instruction about where they should be in the Black Hills and what they should be doing there.

This Hopi horizon calendar was recorded by the anthropologist Alexander M. Stephen in the 1890s. The Sun priest observed the setting positions of the sun behind peaks in Arizona as it approached the winter solstice. When sunset reached the distinctive notch in the horizon where it set on December 10, the priest judged that the traditional nine days of mid-winter ceremony should start in four days.

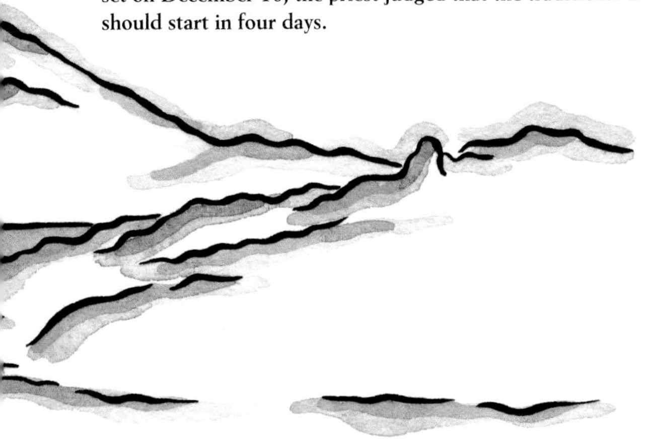

ceremonial calendar, but they are important because they mark these directions and define the time for pilgrimages along them. To the Hopi, there is no clear distinction between space and time; it is more accurate to say that certain places and directions acquire particular significance at certain times, which are defined by natural events such as the sun rising at a certain place.

Ritual calendars of the chiefs

Most examples discussed so far involve small communities or villages forming part of cultural groups related by kinship ties, but with no one group or settlement having dominance over others. Many societies, however, are governed by a chief, and prestige is determined by genealogy—how close the relation to the chief. This can affect the way timekeeping is understood in the society. The chiefdoms that developed in Hawaii during the century or two before European contact are a case in point.

The Hawaiian islands were first inhabited, probably around A.D. 400, by Polynesian people who developed navigational skills to cover vast expanses of the Pacific Ocean in their twin-hulled sailing canoes. Here, where the land and climate were favorable, agriculture was soon established and rapidly intensified as the population thrived. By about A.D. 1650, Hawaiian society was organized into chiefdoms with social ranks ranging from high chiefs (*ali'i*) through lesser chiefs, priests, and other specialists to common fishermen and farmers (there was possibly also a caste of slaves and outcasts).

Polynesian navigators had a good knowledge of the night sky, and elements of a common calendar based upon the same sequence of named lunar months survived on a variety of Pacific islands until European contact. But this was highly developed in Hawaii, with detailed predictions attached to months of the year and to days in the month. This regulated many aspects of ritual activity and daily life, and reinforced chiefly power.

One of the best known features of the Hawaiian calendar, thanks to the work of the anthropologist Marshall Sahlins (b. 1930), is the *makahiki*, a period of

:aptain Cook
1728–1779), the explorer
vho met his death when
e arrived on an island at
time the natives
onsidered inopportune.

rituals, sports, and other activities in honor of Lono, the god of agriculture, which took place at the time of winter rains around the December solstice. According to Sahlins, when Captain James Cook first arrived on the island, which happened to coincide with this period, he was seen as the embodiment of Lono, and treated accordingly with great deference. On his subsequent visit, he arrived after the end of the *makahiki*, which caused great dissonance, and resulted in his death.

Cycles of time in Mesoamerica

If social differentiation develops beyond chiefdoms, the result is usually some form of city state, too large for social ranking to be determined purely by kinship relations. In Mesoamerica—a region of central America covering what is now southern Mexico, Guatemala, and Belize—a succession of great city states developed and collapsed during the two millennia prior to the arrival of the European invaders. This region provides us with valuable evidence about the ways in which timekeeping developed among urban societies that had no conceivable connection with the development of Western ideas in the Old World. The history of Mesoamerica was one of social and political turmoil, yet certain beliefs and practices seem to have persisted, showing remarkable continuity in their development and survival. One of these is the Mesoamerican calendar.

This was made up of two intermeshed cycles: one of 365 days and one of 260. The first of these was relatively straightforward, consisting of eighteen months of twenty days, followed by five "unnamed" days of bad

The first sixty-five days in the 260-day Mesoamerican cycle

1 Imix	1 Ix	1 Manix	1 Ahau	1 Ben
2 Ik	2 Men	2 Lamat	2 Imix	2 Ix
3 Akbal	3 Cib	3 Muluc	3 Ik	3 Men
4 Kan	4 Caban	4 Oc	4 Akbal	4 Cib
5 Chicchan	5 Etznab	5 Chuen	5 Kan	5 Caban
6 Cimi	6 Cauac	6 Eb	6 Chicchan	6 Etznab
7 Manix	7 Ahau	7 Ben	7 Cimi	7 Cauac
8 Lamat	8 Imix	8 Ix	8 Manix	8 Ahau
9 Muluc	9 Ik	9 Men	9 Lamat	9 Imix
10 Oc	10 Akbal	10 Cib	10 Muluc	10 Ik
11 Chuen	11 Kan	11 Caban	11 Oc	11 Akbal
12 Eb	12 Chicchan	12 Etznab	12 Chuen	12 Kan
13 Ben	13 Cimi	13 Cauac	13 Eb	13 Chicchan

ɩesoamerican timekeeping:
Xochicalco relief sculpture
presenting the Nahua and
ıpotec calendars.

omen; but it is the 260-day count that distinguishes the Mesoamerican calendar from any other. In the sense that it runs continuously and independently of the annual cycle it is similar to our week, but there all resemblance ends. The 260 days were identified in what seems to us a most bizarre fashion, by meshing together the numbers from one to thirteen with twenty names: *Imix*, *Ik*, *Akbal*, *Kan*, *Chicchan*, *Cimi*, *Manix*, *Lamat*, *Muluc*, *Oc*, *Chuen*, *Eb*, *Ben*, *Ix*, *Men*, *Cib*, *Caban*, *Etznab*, *Cauac*, and *Ahau*. This means that the day one *Imix* would be followed by two *Ik*, then three *Akbal*, up to thirteen *Ben*, but then the next days would be one *Ix*, two *Men*, up to seven *Ahau*, eight *Imix*, nine *Ik*, etc. Since thirteen and twenty have no common factor, every combination will be covered, and no day name repeated, until 260 days have passed and the whole cycle begins once again.

The origins of the 260-day count and the reasons for its development are unclear, despite much speculation on the subject. It has no obvious basis in astronomical cycles. Yet it was of fundamental importance for divinatory purposes, and was used to regulate a variety of rituals. When the 365- and 260-day cycles are run together, the resulting "Calendar Round" lasts for a full fifty-two years. The end of one of these cycles was a time of great dread, and powerful rites were required in order to ensure the beginning of a new one. These rites included the Aztec "New Fire" ceremony. A priest extinguished flames that were normally kept burning in all the temples. The whole community fasted and prayed, and the more devout members mutilated themselves. When they detected a propitious omen in the heavens, the priests slit open a living human sacrifice and kindled a new sacred fire in his heart. From this one fire, all the temple fires, and household sacred fires, were relit.

Cycles on cycles

The Maya peoples of the Yucatan peninsula are exceptional among the races of Mesoamerica in having developed hieroglyphic writing. Although it was only recognized as such in the second half of the twentieth century, during which time deciphering the code proceeded apace, we now know of many episodes in Mayan history, and aspects of Mayan culture and ritual, that were unknown or misunderstood by earlier scholars.

We are even luckier that, by a few threads of good fortune, some examples have survived of the many bark books, called "codices," that were produced by the Maya, despite the fact that Spanish clerics in the sixteenth century set out to burn all of them. Miraculously, just four are known to have survived, turning up at intervals over the centuries in libraries and private collections, three in Europe and one in Mexico. One of them, the so-called *Dresden Codex*, happens to be an astronomical and astrological almanac.

An astrological calendar at the Inca ruins of Machu Picchu in Peru. Part of an important archeological site, it was discovered in 1911.

It is largely because of the *Dresden Codex* that we know that the Maya became obsessed with interlocking time cycles, over and beyond the two day-cycles already mentioned. We have said little so far about planets, but one of the most influential astronomical bodies for the Maya was the planet Venus. Its appearance in the sky changes distinctly over a cycle lasting 584 days. Sometimes it is a morning star, the brightest object apart from the moon in the sky before dawn, and sometimes it appears in the evening sky after dusk. Twice in its cycle it disappears from view, when it is more or less the same direction as the sun: once for about ninety days, when it is on the far side, and once for eight or nine days, when it is in front of the sun. It was the repeated appearance of the sequence of numbers 236, 90, 250, and 8 that first drew historians' attention to a table occupying five pages of the *Dresden Codex* which turned out to tabulate the motions of Venus to an accuracy of one day in five hundred years. The motivation was primarily astrological. The motions of Venus, for example, were seen as correlated with the coming of rain and the success of the maize crop. Observations could also be used to determine the most propitious time to do things, for example to wage "ritual warfare" in order to capture warriors for sacrifice to the gods.

The ancient Maya recognized a number of what we would see as coincidences of nature as natural rhythms of the cosmos. For example, five Venus cycles amount to almost exactly eight years, and are also very close to ninety-nine lunar cycles. Another coincidence is that forty-six 260-day cycles, which the Maya knew as the *tzolkin*, are almost exactly equal to 405 lunar months. This is the reason that another table in the *Dresden Codex*, which records intervals between '"danger periods" when solar eclipses might occur, covers exactly this period.

The Long Count of the Maya

The Maya not only possessed writing, but also a base-number system, something that distinguished them, for instance, from the Romans. We depict numbers using only the digits 0 to 9 and represent arbitrarily large numbers by using digits in various positions to represent multiples of powers of ten. For example:

$$4{,}362 = 4 \times 100 + 3 \times 100 + 6 \times 10 + 2.$$

How the Maya wrote their numbers

In the Dresden Codex dots represent the number 1, bars the number 5. Places—each increasing in power by 20, instead of our 10—are read vertically, so that the number highlighted in blue opposite, on one of two pages from the Codex, is equivalent to 20 × (5 + 3) + (5 + 5 + 5 +2), in other words 160 + 17, which equals 177.

A representation of extracts from the *Dresden Codex,* which helped unlock the secrets of Mayan astronomy.

In a similar way, the Mayans used dot-and-bar symbols to represent "digits" from 0 to 19, and expressed numbers using powers of 20. The ability to represent large numbers easily was undoubtedly a critical factor in helping them to develop—alongside all the cyclical calendars and counts we have already mentioned—the so-called Long Count, a linear calendar that simply counted the numbers of days back to a zero point at what was seen as the creation of the universe. This date corresponds in our Gregorian calendar to about August 13, 3114 B.C. There were no Mayans around at that time; the mythical history that accompanies the Long Count was simply an invention.

In conceptualizing this history using the Long Count, the Mayans devised a scale of divisions and subdivisions of time-cycles and subcycles. Each day can be expressed in terms of the number of *baktuns* (periods of four hundred years), *katuns* (twenty years), *tuns* ("years" of 360 days), *uinals* (twenty days) and *kins* (days) since the creation, something that (apart from the *tuns*) is equivalent to the number of days written in base 20.

If you are reading this chapter before June 5, 2012 you will not yet have experienced the most significant event predicted in the Maya Long Count—but it is important, for this is when the Maya calculated that the world would end.

A sobering thought is that without their writing we would know little or nothing of the existence—let alone the distinctiveness or complexities—of Mayan astronomy. Although there has been speculation that a few rare circular buildings built by the Maya may have been observatories, there is no convincing archaeological evidence of any observing instruments.

Obsessed by Western ideas

In the modern Western world we tend to assume that our way of understanding the universe in which we live and our place within it, rooted in the ideas and methods that have accumulated as part of the Western scientific tradition, is essentially the correct one. We certainly feel we have developed the right general approaches and methods to ensure that we will continually improve upon our knowledge through theory and experiment in the future. While there is no need for any anthropologist, historian, or archaeologist brought up in the Western tradition to question this point of view, anyone studying human societies situated outside that tradition, past or present, must beware of assuming that there is an inevitable progression of ideas on anything that leads to our present state of knowledge.

From an anthropological point of view, one of the main dangers in doing this is to judge other societies against our own, measuring their progress against the yardstick of our own achievements. Apart from the problem of how patronizing this must seem, for example, to modern indigenous communities, it denies that other systems of thought can be perfectly self-consistent and valid in their own terms, and that they can have great value for the society concerned. We should not close our eyes to the rich variety of ways in which human communities have contrived to understand the world around them and, on the basis of that understanding, to determine the best ways of keeping their activities in tune with their environment. Although the frameworks of thought involved are often quite unlike those holding sway in the Western world, they have helped human societies thrive in a great variety of circumstances.

Nowhere is it more tempting to assume an inevitable progression of human thought than when discussing the development of calendars. Any attempt to identify the main steps along the way can be countered by rich examples at almost every stage. Each one bears witness to human wit and diversity and we are the richer and wiser for knowing about them.

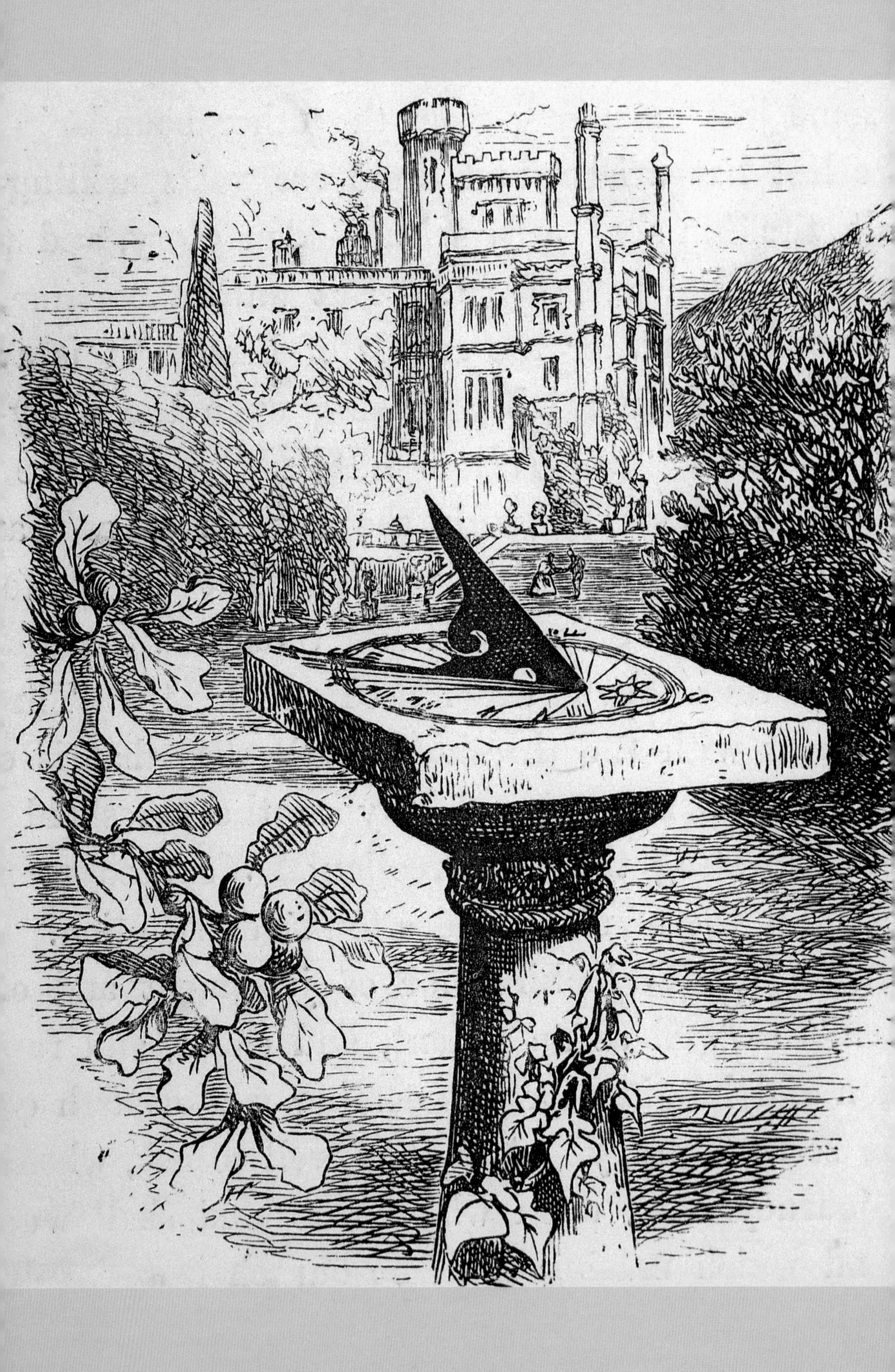

6: the time of day

Sara Schechner

Marking the sun's passing

The gods confound the man
who first found out
How to distinguish hours.
Confound him, too,
Who in this place set up
a sundial,
To cut and hack my days
so wretchedly
Into small portions! Plautus

From ancient times until the modern era, the sun and stars were used to find time by day or night. Specialized time-finding instruments were invented for this purpose. These included sundials, astrolabes, quadrants, nocturnals, and star clocks. Until the eighteenth century, sundials were the most common and accurate method for finding time during the day. Even people who owned clocks required sundials to set them. Many fixed sundials survive in their original settings, while thousands more survive in museum collections worldwide. From them, we learn not only how people found the time, but also how their attitudes toward, and relationships with, time have changed.

Early sundials

Sundials came in many shapes and forms, but common to all was a *gnomon*. The word "gnomon" comes from the Greek, meaning "to show or indicate." A *gnomon* was any object that cast a shadow whose length or position was used to indicate time or geographical direction. In a sundial, the *gnomon*'s shadow fell on a calibrated surface marked with hour lines. They were among the earliest astronomical instruments. By means of shadows, they marked the passage of time, the recurring seasons, and the apparent motion of the sun in the sky. The oldest surviving examples are Egyptian and date from about 1500 B.C. Nevertheless, sundials were probably introduced much earlier in the third millennium when the Egyptians began to divide the night and day into twelve seasonal hours each. The twenty-four-hour day and the art of making sundials reached ancient Greece via the Babylonians. Anaximander of Miletus is credited with setting up a *gnomon* in Sparta in the sixth century B.C. Literary

Nighttime hours in Egypt

As we saw in chapter four, the Egyptians recognized thirty-six time-telling stars (the *decans*), whose morning risings were linked at one time to the thirty-six administrative weeks of the year. However, besides the *decan* rising just before dawn on any given night, about eleven others could be observed before it, rising one after the other.

Because of the twilight after sunset and before dawn, only about twelve of the eighteen *decans* passing overhead in half a day would be clearly visible. So the night was divided into twelve equal parts, each associated with the rising of a *decan*. The twelvefold division of the night led the Egyptians to divide the daylight period into twelve equal parts as well.

Since night and day always had twelve hours apiece, the lengths of the hours varied with the seasons. Daylight hours shrank in duration during the winter and swelled during the summer, while nighttime hours did just the opposite.

lamic astronomers of e 1500s at work in the tanbul observatory. At e upper right two tronomers examine an trolabe, which could e used for telling the me by fixing the osition of the sun. The tronomer top center is xing the sun's position ith a quadrant. There an hour sandglass and mechanical clock niddle right).

references to these early sundials indicate that their functions were primarily astronomical and calendrical, and that they were used to mark solstices and equinoxes.

Hour-recording sundials did not become common in Greece and Rome until the third century B.C. But then they became very common. Sundials were placed in private courtyards and in public squares, near temples, and in public baths. Used to coordinate meal times and other activities, they became social tools for structuring the day.

How did this happen? How did sundials become transformed from astronomical instruments into tools for daily life?

Time consciousness and time discipline

To begin to answer these questions, we must ask two more. Would a common person in the ancient world—someone who was neither priest nor philosopher—own a sundial? What was his or her sense of time?

The answers depend in part on whether the setting was rural or urban. For people living in a rural society, sundials can have served little function. The daily sequence was defined less by social conventions than by the rhythms of agriculture, the cycle of the seasons, and the rising and setting of the sun. Recurring patterns of shadows that sweep across the landscape were part of this natural cycle too. By keeping track of mowing and sowing—or of shadow lengths and positions (whether of one's own body, a stick, or a sundial)—one enters into a world of time consciousness.

Time consciousness itself has varied with time and place. In traditional agricultural societies, the sense of time was cyclical. Time was divided into recurring units—days, seasons, cycles of birth and death, and regular bodily urges. In ancient Greece and Rome, for instance, the common people partitioned the day into three or four segments. These were given names like *gallicantus* (time of the cock's crow), or names derived from routine activities or meal times. "The most accurate clock in the world is the peasant's stomach," wrote Tommaso Garzoni in 1586, expressing a long-cherished opinion.

But there were those who felt that a peasant's belly was not a timepiece on which you could depend. A Roman comic writer of the late third century B.C., perhaps Plautus, complained about them in these terms:

The gods confound the man who first found out
How to distinguish hours. Confound him, too,
Who in this place set up a sundial,
To cut and hack my days so wretchedly
Into small pieces! When I was a boy,
My belly was my sundial—one surer,
Truer, and more exact than any of them.
This dial told me when 'twas proper time
To go to dinner, when I ought to eat;
But nowadays, why even when I have [plenty],
I can't fall to [i.e., dig in] unless the sun gives leave.
The town's so full of these confounded dials.

As this hungry author noted, the introduction of sundials not only reinforced time consciousness, but also encouraged time discipline. In the third century B.C., this was a new urban experience.

Since they did not have the same natural rhythm of chores, city dwellers of the ancient world were more interested than their rural counterparts in scheduling their days according to a dial. This scheduling was seen variously as unnatural, humorous, or praiseworthy. But timekeeping was not confined to the lives of busy city folk; perhaps the most disciplined time-finder was the Christian Church.

Church time

The Church adopted the Roman system of unequal, temporal hours and structured the liturgy around it. In the Roman system, derived from the Babylonians and Egyptians, there were twelve daylight hours and twelve nighttime hours. The relative lengths of these hours varied with the seasons. By definition, the sixth hour was at midday. Church time—like agrarian time—was essentially cyclical. It was built on the recurrence of holy days and the rhythms of prayer.

The theologian Tertullian (160–220) recommended that daily prayers should be scheduled at the third, sixth, or ninth hours of the day, as well as in the morning and evening. The times were set as reminders of the Passion of Jesus. With the rise of monasticism, prayer times became strictly regulated in each monastic house. Most influential was the Rule of Saint Benedict (established c. 530). According to this rule, there were to be seven daytime services and one at night. These were known as offices—or collectively as the Divine Office. The offices of *Prime*, *Terce*, *Sext*, and *None* took their names from the temporal hours with which they were associated. The monks believed that the power of prayer was strengthened when the whole religious community raised their voices

sundial on a church orch roof in England.

collectively. To this end, the Church was more concerned with the daily cycles and rhythms of prayer than it was with the precise moments during the day when the prayers were collectively said. As a result, the times of prayer services drifted and moved forward during the day. For instance, the ninth hour service of *None* moved to the sixth or midday hour, giving us the English name of noon for midday. The times of the offices became known as the canonical hours to distinguish them from the civil temporal hours. Eventually, each office became synonymous with its corresponding canonical hour. It was in this sense that the monks were said to "recite the hours."

The offices (prayer services) according to the Benedictine Rule

Lauds (daybreak)

Prime (sunrise)

Terce

Sext

None

Vespers (sunset)

Compline (nightfall)

Vigils (after midnight, eventually merged with Matins)

Matins (before daybreak)

Beyond the monastery

In 606, Pope Sabinian is said to have issued a decree that sundials should be placed on churches to regulate the times for *Terce*, *Sext*, and *None*. Simple scratch dials—sometimes called mass dials—can still be found on hundreds of churches in England and Europe dating from the seventh century. The dials were usually placed on the south wall or alongside the priest's door. Often the canonical hours on these dials were marked by crosses. On later examples—after the canonical hours had shifted forward in the day—the crosses were replaced with the initials T, S, N, V, and C, standing for the names of the divine offices. On the great cathedrals of Europe, sundials were also placed in the hands of angels who could then watch over the hours.

Such fixed sundials did more than help the clergy determine the times for Mass; they were inscribed in suitably prominent positions above church doorways to remind passersby to stop and pray. These were public sundials. They give us a glimpse of a community's relationship to time-finding and religion.

The time-finder in the mosque

According to the Koran and sayings of the prophet Muhammad in the early seventh century, the devout Muslim is to pray five times a day in the direction of Mecca. The prescribed times for prayer are sunset, nightfall, daybreak, midday, and mid-afternoon. Since the eighth century, these times have been defined astronomically and by means of the lengths of shadows. In calling the community to prayer from a minaret of the mosque, the *muezzin* either drew upon folk astronomy and simple shadow methods that were codified in religious law, or he took advice from a local astronomer equipped with mathematical tables or an observing instrument, such as an astrolabe or sundial. Beginning in the thirteenth century, the office of mosque astronomer *(muwaqqit)* was established in Egypt and spread throughout the Islamic world. Sundials calibrated with prayer times as well as hours were common equipment in medieval mosques. Islamic astrolabes also frequently have prayer lines.

Sundials could also be private property, serving the personal needs of the devout. In Chaucer's *Canterbury Tales*, a gentle monk consults his portable cylinder dial before inviting the wife of his host to mass and breakfast,

> *"Goth now youre wey," quod he, "al stille and softe,*
> *And lat us dyne as soone as that ye may;*
> *For by my chilyndre it is pryme of day."* The Shipman's Tale

Admittedly, this monk and the good wife were up before Prime planning a tryst, so this incident does not show the most reverent use of a sundial.

Perhaps the famous Saxon sundial unearthed in recent times in the cloister courtyard at Canterbury Cathedral had a more God-fearing owner. This tenth century, bejeweled, altitude sundial emphasized the liturgical times for *Terce*, *Sext*, and *None*. On more certain ground, an ivory diptych pocket sundial adorned with a prelate's hat and the arms of the Colonna family was undoubtedly made for Cardinal Pompeo Colonna (1479–1532) or Cardinal Ascanio Colonna (1560–1608).

Imagery and piety

Many sundials made for priests or devout laity survive in museum collections. They were ornamented with scenes from the life of Jesus—including the nativity, flagellation, crucifixion, and resurrection.

The Virgin, saints, and contemporary Church dignitaries were also portrayed. Biblical scenes were popular. Other sundials had religious

symbols such as the monogram of Jesus (IHS), a pierced heart, the Cross, or a dove. Sometimes the *gnomons* were shaped in the form of these symbols. Others had pious mottos such as "Maria help us." Sundials served the devout in many ways as well as helping them to find the times for prayer. Iconography reinforced piety.

Shape was another means of giving voice to one's faith. Christoph Clavius (1538–1612), the renowned Jesuit astronomer, as well as mathematicians in other religious orders, advised the faithful to erect sundials shaped as crosses. The arms of the cross were inscribed with hour lines, and the edges served as *gnomons*. Large stone crosses sometimes marked parish boundaries or were set in church graveyards. What better symbol to remind wayfarers to take time to live righteously? Small brass or ivory crosses were hung on cords around the neck, suspended from a belt or rosary, or carried in the pocket. They also doubled up as amulets to ward off evil spirits.

The finest cruciform sundials also served as reliquaries. In the Adler Planetarium and Astronomy Museum in Chicago, a German example contains tiny bones from four saints and has a little crucifix made of wood allegedly from the Cross, bound in gold. The gilt outer surfaces are adorned with images of the Virgin and a bishop. To serve religion, sundials were inserted inside the covers of breviaries (containing the service of the day) and psalters (a copy of the psalms). Small, book-shaped sundials made in Nuremberg in the sixteenth century called these prayer books to mind. These could also be worn on cords around the neck. In cruciform and book-shaped dials, form and function were well married.

Serving the pilgrim

Many sundials were designed to be portable and work at different latitudes. Known as universal sundials, they were intended for merchants, pilgrims, and other travelers. To determine the right time, the traveler would consult a gazetteer inscribed on the instrument or pasted inside its case. The gazetteer listed cities (and occasionally holy shrines) along with their latitudes. The traveler would use the information to incline the sundial at the proper angle.

The Adler Planetarium owns a remarkable sundial belonging to a pilgrim. It is a two-leaved, wooden compendium made by Erhard Etzlaub of Nuremberg in 1513. In place of the gazetteer, Etzlaub cut a map into the top leaf to show pilgrims the way to Rome, Jerusalem, and Sinai. If a pilgrim were unsure of his location, he could use the quadrant on the lower leaf in order to find his latitude from the sun or stars. Once latitude was determined from the map or quadrant, the pilgrim opened up the instrument and used the corresponding hour scale. There even was a special hour scale for use in Sinai or Egypt.

fourteenth-century gel cradles a sixteenth-ntury sundial on artres Cathedral in ance. Sundials served ristianity through their nctions, shapes, and namentation. They re used to find the urs of the Divine fice, the dates of ster and other festivals, d the way to Rome and rusalem. They served grims, held the bones saints, and encouraged aceful meditation.

Portable dials from Etzlaub's workshop—like those from many other shops—indicated the lengths of day and night so that hours for travel could be determined. It was common procedure for sixteenth- and seventeenth-century universal sundials to find the time simultaneously in Nuremberg hours, Italian hours, and common hours so that the traveler would not be caught off-guard when crossing a border. Sundials of the type made by Charles Bloud and Jacques Senecal in mid-seventeenth-century Dieppe, France, went to greater lengths to accommodate travelers' needs. Some had a handy *Guide Michelin* built in—a table of information about the services in each town, including the highest ranking ecclesiastical authority or noble, schools, courts, inns, stables for post horses, and public baths.

Eastertime

In addition to setting the times for prayer, the Church was concerned to set the date of Easter each year and the cycle of movable feast days that were coupled to it. Since Easter was celebrated on the first Sunday after the first full moon of the vernal equinox, it was tied to both the solar and lunar cycles. Methods for computing future dates of Easter brought the theologian and astronomer together. Christoph Clavius, the Jesuit

A pocket-sized sundia marked with hours an minutes. This was constructed in 1625, Dieppe in France.

astronomer and author of many treatises on the mathematics of time-finding, was selected by Pope Gregory to reform the calendar to ensure that Easter fell in the right season. The Gregorian calendar was inaugurated in 1581.

The science of scheduling Easter was beyond the grasp of the average person, but the well-to-do had only to reach for their sundials to hide this fact. During the Renaissance, the finer instruments had volvelles (rotating disks with pointers) or tables that could be used to determine the annual dates of Easter and the movable feasts. Thus science came to the service of religion.

Religion in the service of science

Not only did time-finding serve religion, but religion served time-finding. Astronomers were permitted to transform great cathedrals into monumental sundials and astronomical instruments. In 1475, Paolo Toscanelli made a hole in the cupola of Santa Maria del Fiore in Florence to serve as a *gnomon*. A shaft of light entered the church through the hole and descended about three hundred feet to the pavement, where the light fell on a calibrated meridian (north-south) line. In 1574, Egnazio Danti began construction of a meridian line in Santa Maria Novella in Florence, before relocating to Bologna where in 1575 he laid out a meridian in the basilica of San Petronio. In 1579, he traced another in the Vatican's Torre dei Venti. In 1665, Gian Domenico Cassini—chair of astronomy at the University of Bologna—created a much larger version of the San Petronio meridian. The primary function of these instruments was to determine the exact dates of the solstices and equinoxes with a view to reforming the calendar. Cassini also hoped to find the apparent diameter of the sun during the course of the year and from this the eccentricity of the earth's orbit.

Merchant time

From the fourteenth to seventeenth centuries, as feudal society built on the rhythms of the countryside gave way to more urban, commercial society, time became a precious commodity to be budgeted and spent wisely. Agrarian time and monastic time had flowed endlessly in cycles repeated each day and every year. By contrast, merchant time was not the cyclical flow familiar to farmers and friars, but money slipping through the fingers. Time wasted meant missed opportunities and lost profits.

Scholars also felt the strain. Educational reformers urged them to organize or double-up studies methodically to maximize learning in limited hours. Otherwise, they would run the risk of leaving nothing to their credit when they died. Petrarch spoke for many when he noted, "Everything consists in the ordered disposition of time."

This eighteenth-centu work shows Father Time with a sundial. stands on the terrace Duncombe Park, nor England.

Although the reasons for these changes have been much debated within recent years, the fact that change took place is undisputed. With the new time pressures came new images of Chronos or Time. Classically portrayed as a joyful, winged youth holding a sundial, Time came to be seen in the Renaissance as a ruthless old man, an inescapable force causing ruin and decay. Again and again, illustrators and poets depicted Time as a decrepit elder with wings to symbolize his fleeting nature, sandglass to measure time, and scythe to cut down anything in his path. In numerous Renaissance works, Time and Death gloated over their partnership in laying waste to youth, human arts, technology, learning, and power. By the early modern period, Time had acquired lethal instruments and people saw themselves in a battle against him.

Sundials reflected the new attitude toward time. Father Time and other *mementi mori* appeared on sundials to caution people to use time wisely. For instance, two ivory diptych sundials by Hans Troschel and Paul Reinmann, both working in Nuremberg in the late sixteenth century, show a reclining infant with an arm resting on a skull. In the Troschel example, the infant also holds a sandglass, and a motto reads, "*hora fugit mors venit*" (the hour flies; death comes).

The message was still compelling 150 years later, when an artist painted a stained glass window sundial, currently in the Basel Historical Museum. On it, Chronos supports the fabric of time—a cloth marked with the hours—as a skeleton peeks around the edge. And one hundred years later, early nineteenth-century cube sundials by E.C. Stockert and followers depicted Father Time on the western side of the instrument. The western side was the afternoon side, so as the day drew to a close, the user would be reminded of the limits of life and the final reckoning.

Time is money

The science of gnomonics and art of sundials not only contributed to the salvation of souls but also preserved the market economy and society. As the English mathematician William Leybourn declared in 1659,

> *What is more necessary in a well ordered Common-wealth [than dialing]? what action can be performed in due season without it? or what man can appoint any business with another, and not prefix a time, without the losse of that which cannot be re-gained, and ought therefore to most be prized?*

Merchant time was not just linear, but money-conscious. In 1748, the American scientist and revolutionary Benjamin Franklin counseled,

> *Remember, that time is money. He, that can earn ten shillings a day by his labor, and goes abroad, or sits idle one half of that day, though he spends but sixpence during his diversion or idleness, ought not to reckon that the only expense; he has really spent, or rather thrown away, five shillings besides.*

Franklin wiled away some slow hours during sittings of the Continental Congress by designing money for the United States. In 1787, the first currency authorized by the new country was a one cent coin—the penny. Franklin's design showed the sun in splendor above a horizontal sundial with the motto "FUGIO" (I fly). Below the dial, the coin read, "MIND YOUR BUSINESS." The so-called Fugio motif also appeared on dollar coins and on some paper notes.

Clocks and sundials

It is often said that the new linear sense of time was clock-driven, but the development of clocks was more a consequence than a cause of the sense of urgency that arose in the late Middle Ages and Renaissance. Clocks were seen as tools to aid the administration of civic life and their bells were used to synchronize work schedules, but the same roles were given to sundials, sandglasses, and calendars. Parallel improvements in these time-finding instruments were symptomatic of the time pressures people had begun to feel.

Clocks did not put sundials out of business. In the fifteenth century, sundials appeared that were calibrated for equal hours, so that clocks and sundials could use the same system. These sundials divided the hours into half and quarter intervals. Contemporary clocks were only just being equipped with an hour hand and dial. It was another century before they recorded half-hour intervals. Clocks were notoriously inaccurate as well; their owners had to use a sundial to set them to the correct time. Sundials regulated clocks.

The rise of the clockmaking profession was driven by demands from royal houses and very wealthy merchants for personal timepieces. These domestic clocks were beyond the means of individuals and institutions with modest finances. Sundials filled the gaps. For instance, the Ecclesiastical Constitutions of Saxony and Lower Saxony, passed in 1580 and 1585, called for sundials in all parishes that had no clock:

> *And in villages without a clock, the pastor should admonish the church, and in particular the people who can afford it, to buy one, so that the church-offices can be carried out at the appropriate time in accord with the clock, and the people in other respects, too, should be guided by it in their housekeeping.*
>
> *But if the parishioners are so poor that they cannot buy a striking clock, the pastor shall give thought to a sundial, which can be obtained at little cost. And until it is installed, the sextons shall learn from the pastor who has a compass [sundial], or purchase it themselves, and the sexton shall use it to determine the ringing.*

Clocks and sundials were developed in parallel and became more diffused throughout society. Improvements in both were manifestations of the need people felt to use time wisely or impress their associates.

Sundials and consumer culture

Pressure to schedule life events encouraged a dramatic increase in the production of sundials, in their availability to all levels of society, and in their increased precision. The necessity of sundials was seen to be so obvious that it hardly needed comment. "Here would be the place to report all the uses of sundials," Philippe de la Hire wrote in 1698. "But I believe it suffices to say that there is no state in life in which exact knowledge of the time is unwarranted, be it for regulating matters of religion and the Divine Office, or for apportioning work and rest fairly."

But missed appointments and worries about wasting time do not fully explain the increased demand for sundials, the types bought, or their meaning in people's lives. The art of dialing was part of polite culture. Many authors commended it as a "rational and instructive Amusement to the young Student...and to Ladies and Gentlemen in general." A competent understanding of its first principles and the relationship of gnomonics to astronomy, geometry, and optics was declared "requisite for forming the *Scholar* and the *Gentleman*" and essential for "making a man compleat [sic] and excellent." Under such circumstances, the possession of a sundial was a sign of good breeding and a marker of some status. Without doubt, many owners did not understand the mathematical projections underlying their pocket sundials, but they could pretend they did or declare that they had the means to learn the mathematics if they ever deigned to do so. The more complex their instruments were, the greater the prestige they bestowed on their owners.

Many types of sundial provide evidence of conspicuous consumption while expressing their owner's taste for mathematically sophisticated and complex instrumentation. Very often such instruments were commissioned. An astronomical compendium in the Alder Planetarium's collection provides a fine example. It is dated 1557 and signed "V.C." (most likely a Flemish artisan employed by Thomas Gemini in London). Compendia were the "Swiss army knives" of the sundial world. They combined time-finding instruments into a pocket-sized package. The V.C. compendium consists of thirteen leaves hinged together in the form of a book. The leaves contain astronomical and astrological tables, lunar volvelles, an aspectarium, calendrical and horological tables, a horizontal sundial, equatorial sundials, moon dials, a quadrant, a nocturnal, star charts, and a world map. Compendia were not the only form of showy sundial, another piece in the Alder Planetarium's collection—a gilt brass and silver pillar dial—that gives new meaning to the phrase "conspicuous consumption." Probably made in Prague in the 1580s, the pillar housed a retractable knife and a fork. The user could check his sundial, declare it time for dinner, and after that he could convert his impressive instrument into his eating utensils!

Sundials of a single type often display a range of styles (simple to ornate), materials (cheap to precious), or workmanship (crude to masterful). Such cases suggest that the given type appealed both in function and form to people up and down the social ladder. Both the sundials and the makers' catalogs inform us that makers adapted their wares to fit the special needs of both elite and common consumers.

Instruments at the top end of the market were made of silver and gilt brass. At the low end they were constructed cheaply of printed paper glued to wood. Michael Butterfield, an English instrument-maker working in Paris between 1665 and 1724, created a cute and popular style of horizontal, pocket sundial in which the *gnomon* was shaped as a little bird. Butterfield made his best sundials of silver rather than brass and marked them *premier cadran* (top of the line). Bion, Baradelle, Haye, le Maire, and other makers of Butterfield-type sundials likewise divided consumers by metals. The same story of high and low markets can be told with the portable, equatorial sundials known as Augsburg-type sundials.

Gender differences are also observable. In the Alder Planetarium's collection, two instruments illustrate this point. One is a round Augsburg-type dial constructed of silver and gilt brass by Johann Martin of Augsburg, c. 1675–1700. The sundial folds up and fits into a silver box on which there is a perpetual calendar volvelle. The silver box, in turn, is stored in a brass box, and the final package is two and one-half inches in diameter. It is designed to fit in a man's pocket. The second instrument was made by Philip Happacher during the same period. It is similar to the Martin sundial, but is daintier and has a gold case with a perpetual calendar. A mere one inch across, this instrument was suitable for a wealthy lady's purse.

Sundials bear witness to a flow of aristocratic tastes and values down the social ladder. Johann Martin and Johann Mathias Willebrand made Augsburg-type sundials in the late 1600s with a high degree of precision and the finest materials. About a century later the same type of sundial was being produced by Johann Nepomuk Schretteger and Lorenz Grassl with more affordable materials and with less precision. Evidence such as this points to the lower orders choosing to emulate their social superiors as upper class style and technology became vulgarized over a period of roughly one hundred years. Elite consumers often spurned the older styles as they became too common.

Some sundial-makers were imposters who pirated copies of trendy styles or counterfeited the dials of famous makers. Forgeries satisfied people's desire to enjoy a status item at a discount, unless they were taken in by a swindler. They tell us much about the popularity of particular styles and instrument-makers. Makers would not forge an undesirable instrument, and so sundials falsely attributed to Butterfield that survive in museum collections tell us that his instruments were in demand.

sixteenth- or
venteeth-century
cket sundial. It lets
nlight through a
all hole to reach the
l on the rim.

Where a sundial was used

The cities inscribed on sundials were listed in gazetteers. This tells us about the geographical interests (if not the actual destinations) of some consumers. For instance, sundials made in Paris in the mid-eighteenth century show not only a French perspective of Europe, but also an interest in French possessions overseas. The French government supplied dials like these to its officers in North America from 1751 to 1759 "to guide them through the forests and wastes of that country." And the French were not alone in catering to foreign markets. German, Italian, and British instrument-makers also created instruments for sale or use abroad.

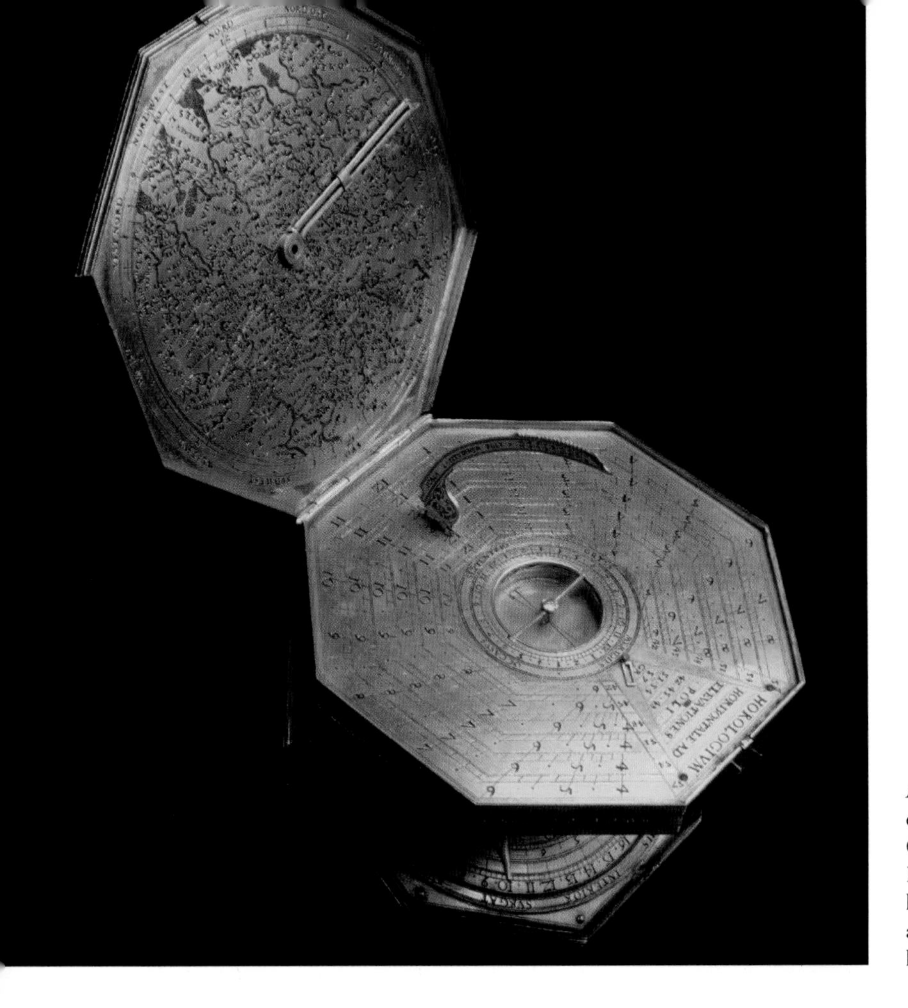

An astronomical compendium made in Germany in 1557. It consists of thirteen leaves of tables, charts, and dials, hinged togeth like a book.

However, the fact that a gazetteer lists North American territories does not imply that the dial-owner actually ever left Europe. A lot of armchair traveling was done in these times, and sundials enabled their owners to fantasize about journeys to far-flung places like Quebec or Calcutta. It is unlikely that the owners of many pocket sundials ever traveled as widely as the gazetteers suggest. All that can be said for certain about such dials is that the owner thought of himself, or wanted others to think of him, as the sort of person requiring all the extra information.

Specialized scales can offer more certain information about the location of their users. Most Butterfield-type sundials made by Baradelle of Paris (c. 1752–94) were inscribed with hour scales for forty, forty-five, forty-nine, and fifty-two degrees of latitude, which suited European needs. Although one surviving instrument has hour scales for eighteen, twenty, and twenty-five degrees of latitude, which is too low for Europe, it is marked for Santo Domingo and was undoubtedly made for a customer in the Caribbean. Sundials made in the French coastal town of Dieppe frequently included volvelles for determining the times of high and low tides. Tide volvelles were not included on those instruments destined for inland users.

portable sundial, of the
pe that once found
eat popularity for
ne-telling.

Precision

The physical construction of sundials sheds light not only on the skill of their makers, but also on the precise nature of time-telling wanted by consumers. Among the sundials made by Michael Butterfield in Paris in about 1700, the Alder Planetarium has a large silver dial marked *premier cadran* with five-hour scales graduated to quarter hours, thirty cities on the gazetteer, and thirty-two points on the magnetic compass. Another silver sundial in the collection is much smaller and has only three-hour scales of which two are graduated to half hours, an eighteen-city gazetteer, and a four-point compass. This smaller dial was more appropriate for a well-heeled schoolboy than a gentleman; it belonged to Brook Taylor, the mathematician, when he was just sixteen years old and entering Cambridge.

Butterfield chose to divide the latitude scales of his instruments into one-degree intervals, whereas Edmund Culpeper, his contemporary, felt that two-degree intervals were good enough. Culpeper, however, divided his hour scale into five-minute intervals, rather than the fifteen- to thirty-minute segments found on Butterfield's dials. Precision in latitude measurement was apparently less important to Culpeper's clients than the prospect of precision in marking time.

7: beyond dead reckoning

Stuart McCready

Finding latitude and longitude without clocks

They that go down to the Sea in Ships, that do business in great waters, these see the Works of the Lord, And His wonders in the deep Psalm 107

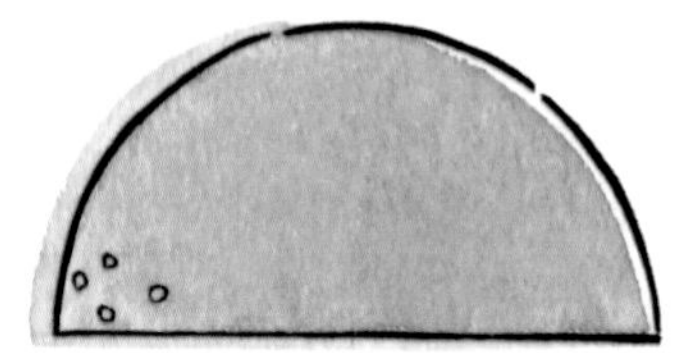

The Southern Cross rising

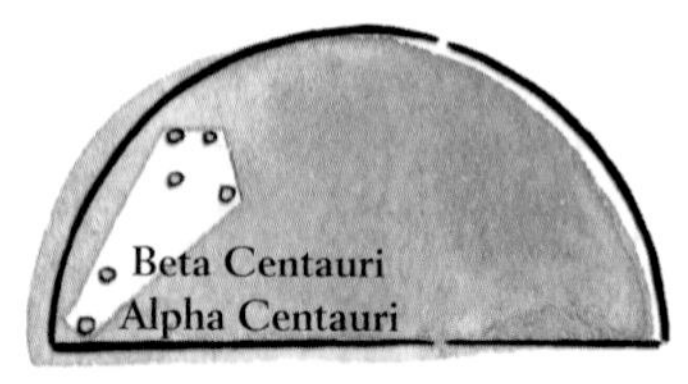

The Southern Cross when Alpha Centauri is rising

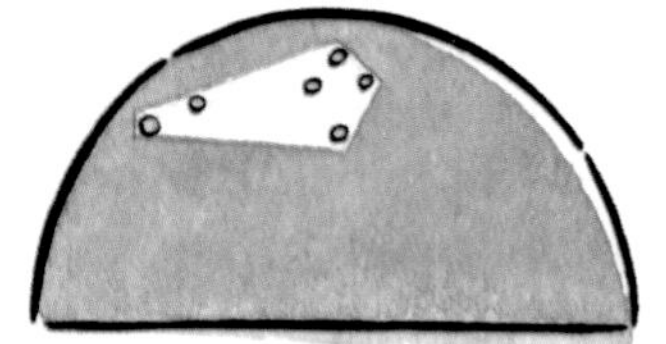

The Southern Cross upright (pointing south)

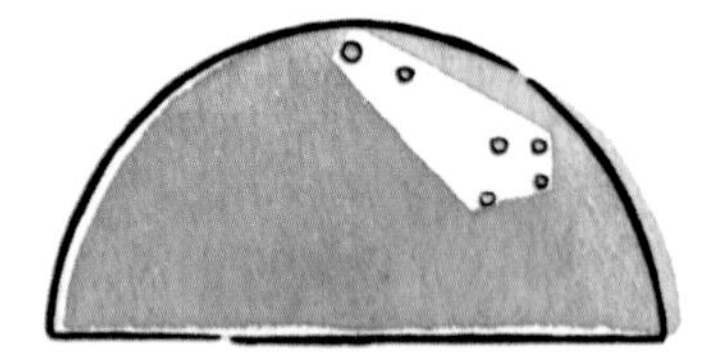

The Southern Cross when Alpha Centauri is at its highest point

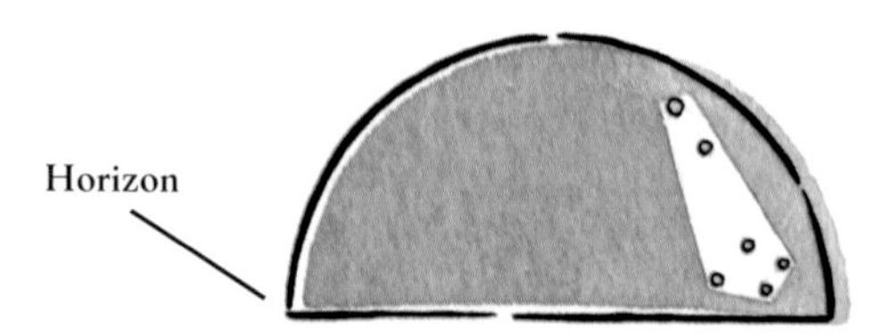

The Southern Cross setting

Using the night sky as a compass

On clear nights, mariners in the Northern Hemisphere can use the near-stationary Pole Star to find north. In ancient times they probably used rising and setting stars as well to find other directions. The sky compass would be read like a clock face on which most of the pointers moved through the night, and pointed in different directions as the night progressed.

No details of how Europeans pictured such a compass remain. However, until recent times Caroline Islanders found their way across empty expanses of the tropical western Pacific without modern navigational aids. As they sailed from one isolated island to another in their outrigger canoes, they relied on a star compass that has been recorded by anthropologists.

The thirty-two directions of the Caroline Islanders' star compass were defined by the Pole Star in the north, the rising and setting Southern Cross constellation in the south, and the rising and setting points of selected stars and star groups in the skies in between.

The Caroline Islands lie north of the equator; so the Pole Star can be seen just above the northern horizon. It moves in a tight circle only one degree from the celestial north pole, the point in the northern sky around which all the stars appear to revolve. The illusion of the stars revolving is an effect of the rotation of the earth; so the celestial north pole is in line with the axis of the earth's rotation. It lies in the same direction as the geographic north pole, the northern end of the earth's axis of rotation.

The Pole Star is a more accurate indication of north than a magnetic compass, since the needle of the compass is drawn towards the magnetic north pole. At present this lies in the arctic archipelago of Canada, which is about 1,190 miles (2,000 kilometers) from the geographic pole.

No easily observed stationary star marks the southern celestial pole, which the Caroline Islanders, in any event, cannot see. The rotation of the Southern Cross around the pole, however, was sufficient to provide them with direction.

The five positions shown left define the five most southerly directions of the Caroline Islanders' compass. Two positions of the Southern Cross are defined by the position of Alpha Centauri, a prominent star in the constellation of the Centaur.

The Southern Cross constellation, as used by Caroline Islanders for orientation.

vessels overtook some piece of weed, flotsam, or foam frothing on the surface. He would note the time it took the length of the ship to pass the object by. Then, adjusting for changes in speed he had noticed during the day, he would reckon the day's progress.

This style of reckoning usually amounted to judging, in the light of experience, that a certain figure seemed right. Calculation based on measurement was not a genuine possibility. Columbus might have counted the seconds as he watched a piece of flotsam pass by, but he had no way of checking the accuracy of his count because there was no mechanical timepiece on board. Anyway, clocks in 1492 still did not count minutes, let alone seconds. They were mostly elaborate, fixed devices (often without faces) that sounded the hours in cathedral towers on land. Columbus's flagship, the *Santa Maria*, had a sandglass (for telling when it was time for him to say his devotions), but it measured half hours. A ship making twenty or so leagues per day would have left a sample of flotsam out of sight almost half a league behind before the sand ran out.

In his log Columbus confesses to misleading the restless seamen under him, announcing fewer leagues per day than he thought accurate. If the voyage became longer than expected he didn't want them to claim they were sailing on without hope past the point, 750 leagues west of the Canary islands, where he had promised they would see land. In fact, his honest estimates were over-optimistic. The speeds he gave out falsely proved on modern analysis to be closer to the truth.

Counting the knots

The timing of a ship's progress improved in the next century with use of the chip log, a standard piece of equipment attached to a line. The log was dropped into the water at the stern and the length of line that reeled out behind the ship told a navigator, who watched a much quicker sandglass than Columbus would have done, how fast the ship was moving away.

In 1637, Richard Norwood, an English navigator, introduced the practice of knotting the line every forty-seven feet, three inches (14.3 meters) and counting how many knots reeled out before a twenty-eight-second sandglass emptied. This is the origin of measuring nautical speed in knots, which are equivalent to nautical miles per hour. If the first of Norwood's knots appeared in twenty-eight seconds, the ship was then said to be traveling at one knot—or one nautical mile (1.15 statutory miles) per hour.

Just fifty years later the first patent log—with a vaned and metered rotor—was towed from the stern of a ship to count knots that had taken on a purely abstract existence. This system, much refined, is still in use today.

depiction of the fleet
Christopher
olumbus, on their first
yage in 1492.

Reckoning direction: the magnetic turn

Columbus set out on his great adventure without the benefit of these aids to reckoning speed. By his time though, navigators had made dramatic progress in their other task—reckoning direction. In the 1300s and 1400s, the magnetic compass had come into common use. It had revolutionized trade in the Mediterranean, as commercial fleets no longer stayed in port through the winter, when clouds had made it impossible to read the compass of the sky.

A new freedom to sail when and where they liked soon took Mediterranean sailors into uncharted waters. In the 1300s, ships from the Italian port of Genoa used compasses to guide them back to shore as they probed the Atlantic in search of rumored islands with tradable wealth. The Genoese rediscovered the Canaries, where the descendants of the Phoenician colonists had been isolated for almost one thousand years, and went on to Madeira and the Azores. They promised themselves they would find a way past the contrary winds and currents on the coasts of West Africa, go around the southern tip they hoped this continent had, and sail all the way to India.

In the 1400s, Spain and Portugal, determined to conquer the Atlantic for themselves, developed a bigger and better ship, the caravel. It took about thirty men to sail one, the same number as the biggest Viking longboat, but it could carry four times as much cargo and provisions—about forty tons. The Portuguese displaced the Genoese as challengers of the South Atlantic and monopolized the search for a way into the Indian Ocean around Africa. Bartholomew Diaz would eventually find it in 1487, by sailing west into the unknown Atlantic, searching for a wind that would carry him south.

Large Portuguese fleets were soon sailing the southern Atlantic route he blazed. Almost as soon, the Spanish, who had backed the rival scheme of Christopher Columbus—to sail even farther west, seeking not a wind but the East itself—were sending fleets to the Caribbean. For the first time, European sailors found themselves, by their own intention, not tens of leagues but hundreds of leagues from any shoreline landmark or any soundable bottom. Ambition, better ships, and, above all, the compass had made this possible; but distance calculations remained inexact and compasses were not perfect. The compass, a boon to direction reckoning, had in turn created a need for a new system of positional reference that could serve as an independent check.

The new chartmakers

Compass-led navigation brought seaborne commerce in the Mediterranean to life, and the chartmaker's trade thrived, especially in Genoa and Venice. The charts showed distances and "rhumb" lines. Each rhumb line was one of thirty-two compass point bearings radiating out in a thirty-two point "rose" from each major port on the chart. If his course was not already charted by a single rhumb line that conveniently ran from port to port, the ship's pilot would draw a straight line between two ports as parallel as possible to one of the rhumb lines, then follow it by compass.

As the Portuguese ventured south along the coast of Africa, they tried to make charts that extended the Mediterranean system of rhumb lines into these new waters. They soon realized, though, that over these long distances, the curvature of the spherical planet they were sailing upon made straight-line navigation impossible. However, the solution to the chartmaker's problem was at hand—in the spherical geometry at which astronomers and geographers had long been adept. Charts now became grids that represented the great circles of latitude and longitude into which the earth is conventionally divided geometrically.

Besides aiding the task of representing curved surfaces on a flat chart, this gridwork gave navigators an infinite number of reference points. Points of reference on the old charts had been a sprinkling of ports, lighthouses, and other landmarks. You could describe your position only when you were between landmarks, in terms of reckoned distances and directions from them. Now it was possible, in principle, to specify any point on the chart by a system of coordinates that was based ultimately on astronomical relationships. If, in practice, a navigator could identify his latitude and longitude, he would have a life-saving check against his reckonings no matter where his ship might be.

Finding latitude by the Pole Star

The Pole Star has always told nighttime navigators in the Northern Hemisphere which way is north. It was also a clue to whether a ship was making any progress northwards. The farther you move from the equator the higher you have to look in the sky to find the star. In the 1500s navigators began to translate this principle into precise information about how far north or south they were. Astronomical instruments were adapted for shipboard use to measure the height of celestial bodies above the horizon. Astronomical tables showed precisely where the Pole Star stood at each hour of the night in its circuit around the celestial pole. A mariner could therefore calculate his ship's latitude by geometrical methods.

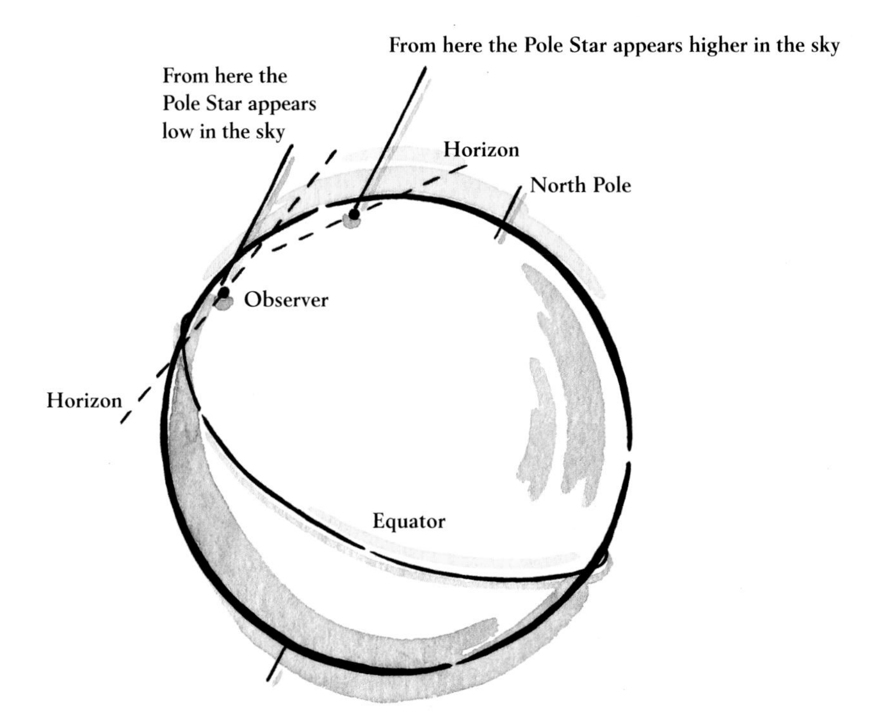

Finding latitude

In practice, it was possible to find latitude (how far north or south you are from the equator), and with increasing accuracy. At the beginning of the 1500s astronomers already had a huge body of observations—building on those of ancient times—of the motions of the most easily observed celestial bodies. These allowed them to predict how high the sun, for example, would appear in the sky from any given latitude at any given time on any particular day. This made it possible—so long as you knew what day it was—to observe how high the sun stood above the horizon at noon, then calculate your latitude.

Astronomers describe how high a celestial body stands in the sky by stating the angle formed when two imaginary lines meet at the observer's eye: one drawn from the celestial body, the other from the horizon. Medieval astronomers measured this angle using a variety of available instruments, such as the quadrant, the astrolabe, or the cross-staff. When Columbus sailed in 1492, astronomical instruments had not yet been adapted for use on the wind-torn, heaving deck of a ship. He used his quadrant only on land, to measure the latitude of Jamaica. During the mid-Atlantic crossing he kept to his latitude entirely by dead reckoning.

The century that followed, however, saw the cross-staff and the astronomer's astrolabe—pared down as much as possible to avoid tossing in the wind—become standard equipment for pilots. Almost every ship carried a printed book of astronomical tables. By the early 1700s the backstaff had replaced the cross-staff. It allowed navigators to compare the position of the sun and the horizon without looking into the sun. The double-reflectant quadrant was invented in 1731 and gradually evolved into the sextant, which today can be used to measure latitude with errors of only a few seconds of arc (a minute of arc is 1/60th of a degree—a second of arc is 1/60th of a minute of arc). This allows calculation to within a few yards of how far a ship is lying north or south of the equator.

Finding latitude at night

Knowing the hour was not a particular problem for a pilot trying to discover his latitude by observing the sun. The conventional time for reading the sun's elevation was at noon, when it told the time itself by standing due south. At night, however, reading the hour was not so straightforward.

The best way of finding latitude at night was to observe the Pole Star. It lies in almost the same direction as the celestial north pole, the still point in the northern sky. The rotation of the earth on its axis makes the rest of the heavens appear to revolve around this point, which is always exactly the same number of degrees above the northern horizon as the observer's latitude. (If you are on the equator—0 degrees of latitude north of it—the celestial north pole is on the northern horizon—0 degrees above it. If you are at the geographic north pole—90 degrees north of the equator—the celestial north pole is directly overhead—90 degrees above the horizon.)

However, the Pole Star is not exactly at the celestial north pole. In the 1500s it was some 3.5 degrees distant, revolving in a tight circle around the pole. In one part of the night, if you relied simply on the Pole Star, it would tell you that you were seven degrees of latitude farther north than it would have told you had you observed it from the same spot several hours earlier or later. Fortunately, astronomers had tabulated the required correction for any particular hour of any particular night, and Europe's newly invented printing industry made the tables available in inexpensive bound volumes.

But how were pilots to tell the hour? Clocks were not accurate and were hard to keep working. The answer was to observe the Pole Star together with one or two stars in the Big Dipper constellation that revolve around the celestial pole in line with it. This line of stars revolves through the sky almost like the hand of a twenty-four–hour clock. The pilot could compare its position with his table and read how many degrees to add to or subtract from the number the Pole Star alone was telling him.

The Pole Star/Big Dipper clock could also be used directly for telling the time, except for one complication. As we saw in Chapter Four, star time (sidereal time) is not the same as sun time. The earth rotates in twenty-three hours and fifty-six minutes, and so the stars revolve around the celestial pole in four minutes less than the length of a solar day. This difference accumulates through the year, and so the Polar Star/Big Dipper clock can tell you the conventional time only with the help of a calendar that plots the stellar clock's slippage in comparison with conventional time.

In the 1500s, an ingenious instrument, called the nocturnal, combined measuring the positions of the Pole Star and its companions with calculating the corrections to find latitude and read the time. The navigator positioned the instrument to sight the Pole Star and a companion star simultaneously, and rotated an inner wheel that was marked into twenty-four hours. When noon aligned with the date, marked on an outer wheel, this brought the mark for the present hour into line with an index bar aligned with the stars. The number of degrees to add or subtract to the altitude of the Pole Star to calculate latitude was marked on the back.

Finding longitude from eclipses of the moon

Astronomical methods for telling longitude were also at hand, but their practicality was much more limited. More than two centuries after Columbus, in September 1707, it was still possible for navigators to be so wrong about their longitude that an entire English fleet, returning home

A medieval astronomer teaches a navigator how to use the cross-staff. In the background, surveyors demonstrate its use in calculating the heights of towers.

from Gibraltar and not knowing its position, collided with well-charted rocks in the Scilly Isles. Four ships sank, with a loss of two thousand lives.

You can calculate your longitude if you know what time it is in two places at once: where you are now and on a north-south line you arbitrarily call 0 degrees of longitude (for a Portuguese sailor of the 1500s, this would have been a line running north–south through Lisbon). The sun passes over 360 degrees of longitude—the entire girth of the earth—every twenty-four hours. That means that any two points that are one hour apart in time are 15 degrees of longitude apart (1/24th of 360 degrees); so if your local time is two hours earlier than the local time at Lisbon, you are 30 degrees of longitude west of the Portuguese capital.

In the days of Columbus, and still by 1707, no clock had been made that could keep time accurately enough at sea to tell a ship's pilot when the hour was striking at home. But several peculiarities of the heavens offered celestial methods for doing this.

One was the fact that an eclipse of the moon is visible simultaneously from wherever it can be seen. The Greek astronomer Hipparchus, who died in about 120 B.C., is the first to have remarked that this would make it possible to know how many hours apart two places are. All you need to do is record the time locally in each place when the eclipse occurs and then compare the two times.

With great excitement, the Venetian explorer Amerigo Vespucci put the principle into practice in 1499. (His published accounts of this and other aspects of his voyages so caught the public imagination that "Amerigo" slipped into common usage, with just a slight alteration, as a name for the entire New World.) Lying at anchor on the coast of what would soon be called South America, he watched an eclipse of the moon, compared his local time with the predicted time of the eclipse in an astronomical almanac for the Italian city of Ferrara, and calculated how many degrees of longitude west of Ferrara he must be.

Similarly, an explorer and an eclipse of the moon helped an English astronomer find the difference in longitude between London and Charleton Island, in James Bay, Canada. In 1633, *The Strange and Dangerous Voyage of Captaine Thomas James*, an account of a search for the Northwest Passage, was published in London. In an appendix, Henry Gellibrand, professor of astronomy at Gresham College, London, calculated the difference in longitude at 79 degrees and 30 minutes of arc. This figure (only 15 minutes of arc in error) could be calculated because Gellibrand and Thomas James arranged, before the captain set off, to note the time when they both observed the predicted eclipse.

Telling time by the moons of Jupiter

Eclipses of the moon are too rare, though, to be of much use in day to day navigation. In addition, the moment of eclipse was hard to define, and this undermined the precision of compared times. That error of 15 minutes of arc in the James/Gellibrand calculation would be enough to put a ship's position in doubt by hundreds of miles.

In 1612, however, Galileo watched through his telescope as one of the four moons of Jupiter that he had discovered two years earlier slipped behind the planet. He went on to observe that such eclipses were much more frequent and much better defined than those of our own satellite. Galileo quickly applied to the King of Spain for a reward on offer since 1598 to whoever could come up with a practical way of finding longitude.

Galileo's method, though, still did not provide a frequent enough check on position to meet the navigator's needs. Focusing a telescope at sea on such tiny celestial objects was at any rate not a practical proposition—even when navigators climbed onto experimental swinging platforms designed to cancel out the rolling of the deck. However, in 1668, when sufficiently accurate tables of the eclipses of the moons of Jupiter became available, astronomers began to use the method on land. They needed to know precisely the differences in longitude between different observatories in order to correlate their observations accurately, and Galileo's method was the best available.

The lunar clock

A third method required even more sophisticated knowledge of the heavens, but ultimately it proved the most practical of them all. This was to read what time it was in Greenwich by comparing the sky with a Greenwich-timed table of distances from the moon of selected stars. (By the time the method became viable, the Greenwich Observatory in London had overtaken Lisbon as the most widely recognized 0 degrees of longitude.)

When, somewhere in the North Atlantic, you find that your local time is noon by comparing the position of the sun with due south, you compare the sun with something local—your due south can't tell you where the sun is for someone else. In telling time by the method of lunar distances, though, mariners compared the position of the moon with the positions of something they shared with observers in Greenwich—the stars.

The moon moves across the sky more swiftly than its background of stars, so at any moment the distance between the moon and any star you might choose as a reference changes. It gives that moment a unique signature. To make it possible for navigators to read this momentary signature, however, was a monumental challenge.

The changing positions of the slowly revolving background stars had to be charted with telescopic precision. (In the mid-1600s, the best star charts were still based on naked-eye observations.) The task did not begin systematically until 1675, when Charles II of England issued a royal warrant appointing John Flamsteed (1646–1719) as the first Astronomer Royal. Flamsteed was charged with "rectifying the tables of the motions

Greenwich Royal Observatory 'belonging to the King's Professor of Astronomy,' showing Greenwich Park in the foreground and ships on the Thames to the left.

of the heavens...so as to find out the so much desired longitude of places for the perfecting of the art of navigation." The astronomer built Greenwich Observatory and began his observations in 1676. His catalog of 3,000 stars, published half a century later, was the first catalog equal to the requirements of the lunar clock.

Tabulating the complex movements of the moon as it wobbles its way around the earth was also an immense undertaking. In 1713, Sir Isaac Newton's *Principia Mathematica* set out the underlying physics that govern the Moon's motions, but errors of detail meant that navigators relying on lunar tables derived from the *Principia* would often have calculated positions that were wrong by hundreds of miles. Using advances in mathematics not available to Newton, though, Tobias Mayer, professor of astronomy in Göttingen, Germany, finally calculated accurate lunar tables in 1754. His widow would eventually receive £3,000 from an appreciative British parliament.

Greenwich Observatory's first annual *Nautical Almanac*, published in 1766—a hundred years after the observatory first took on the task, was a triumph of celestial time-watching in aid of celestial navigation. It was designed specifically to solve the problem of longitude, and it did. Parallel events, however, had by now overtaken the lunar method. Two years earlier, the English clockmaker John Harrison (1693–1776) had already sent "H4," a marine chronometer he'd built as the culmination of a thirty-year effort to build a reliable seagoing timepiece, on a successful trial voyage to Barbados. Celestial navigation reached its zenith just as the clockmakers liberated the human race forever from the need to squint skywards to tell the time.

8: the triumph of the clockmakers

Stuart McCready

Counting the minutes and the seconds

A clock face with the heads of four prophets added in 1443 by Paolo Uccello (1397–1475), from the Duomo, Florence, Italy.

What is this life if,
full of care,
We have no time
to stand and stare? H. D. Davies

Mechanical timekeeping devices have been with us for more than two thousand years. Their history began with geared clockwork instruments in ancient times and they would reach a turning point when cumbersome weight-driven clocks of fourteenth-century Europe began to strike twenty-four more or less equal hours a day. Improvements in clockmaking would eventually make it possible to tell the time to the minute and to the second.

Clockworks of the ancient world

When clocks first appeared, there was nothing new about the idea of clockwork timekeeping. The Greek scientist Archimedes (287–212 B.C.), of the Sicilian city of Syracuse, is said to have invented a clockwork mechanism designed to imitate the relative motions of heavenly bodies. A tiny bronze device, dredged up from the sea near the Greek island of Antikythera, contains thirty-one geared wheels. Made in the first century B.C., it seems to have been a calendric calculator for showing the positions of the sun and the moon in the zodiac, and for calculating the Metonic cycle of nineteen years and 235 lunar months described in Chapter Four.

As early as 1000 B.C., the Egyptians were using non-mechanical water clocks. Water flowed either in or out of a vessel marked with an hour scale. The Greeks and Romans took this idea further by adding automated displays driven and regulated by the water's time-calibrated flow. In the first century B.C., for example, the Macedonian astronomer Andronicus Kyrrhestes built his Tower of Winds in Athens. Roman commentators marveled at its complex assembly of wind vanes and sun dials, and especially its water clock. This drove a planisphere, a circular metal plate inscribed with a map of the heavens. A float attached to a counter-weight by a line twined around a metal shaft rose in the clock as water entered under constant pressure from a spring-fed header tank. As the float rose, the counter-weight turned the shaft. This turned the planisphere through one complete rotation every day. Wires in front of the planisphere served as a dial indicating the hours.

Water clocks of the Middle Ages

References were still being made to geared timekeeping instruments in Byzantine manuscripts of the sixth century A.D. The Islamic world and medieval Europe both carried on the tradition. Through mechanical connections and the efforts of eleven attendants, a magnificent water clock at the east gate of the Great Mosque at Damascus marked each hour with performing bronze statues of falcons, opening and closing doors, bells, and the lighting of a candle at night. The remains of two elaborate medieval water clocks can still be seen at Fez in Morocco. The monks at St. Albans in England had one, in whose water store they filled their buckets during a fire in their abbey in 1198. In the early 1200s, Cologne in Germany had a clockmakers' street—the Urlogengasse.

One of the most remarkable, and probably the most accurate, of water clocks ever made was the thirty-foot-high (nine meters) Heavenly Clockwork completed in A.D. 1088 by the Chinese mandarin Su Sung (1020–1101). It had geared connections to an automatically rotating armillary sphere (circular metal bands set in motion to reproduce the great circles of the heavens, such as the paths of the sun and the moon). There was also a procession of ninety-six figures appearing one after the other in doorways, each to ring its time with bells and gongs.

The fate of Su Sung's Heavenly Clockwork

Su Sung's wonderful clock was astrological. By means of the rotating armillary sphere, it was possible to keep track of phases of the moon independently of cloud-free weather. This provided palace officials with information essential to properly orchestrating the succession of wives and concubines with whom the emperor shared his bed. It was essential that high-ranking wives should conceive princes near the new moon, and that low-ranking concubines should nourish the emperor's Yang (male principle) with their Yin (female principle) near the full moon, when the Yin was strong.

In 1094, a new emperor came to power. Keeping the world in harmony with the heavens was the personal responsibility (and personal claim to legitimacy) of each emperor individually—so, as was customary, a new calendar and a new theoretical underpinning of the emperor's relationship with Heaven were devised. Su Sung's Heavenly Clockwork was thrown out, along with much else from the old regime. By the time Jesuit missionaries arrived in China five centuries later, Su Sung's machine had been utterly forgotten. The emperor Wan-li and all his court were astonished by the automatically ringing bells of the sixteenth-century mechanical clocks (one weight-driven, one spring-driven) that the Westerners brought him as gifts. Yet these were much less elaborate clocks than Su Sung's, and considerably poorer at keeping time.

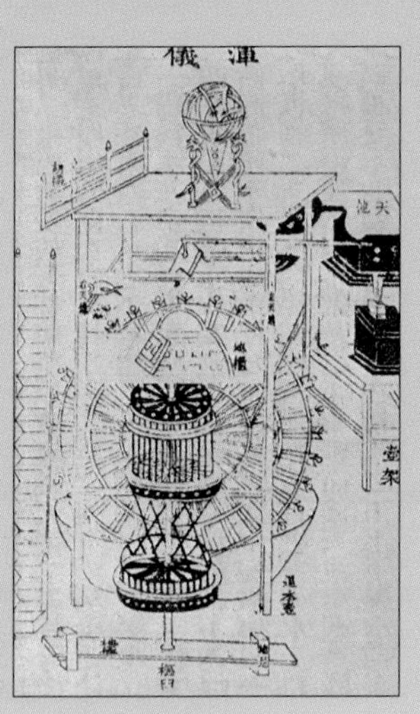

Left: **The design for Su Sung's water clock.**

At the heart of the clock, water poured at a constant rate into scoops attached to a wheel. Every twenty-four seconds, as the wheel stood stationary, a scoop would become just heavy enough with the water that was filling it to fall into a "down" position. This made the wheel rotate by one spoke and brought an empty cup freshly into position under the water inflow. The wheel had thirty-six spokes and so rotated every fourteen minutes and twenty-four seconds—one hundred times a day in all. This is the first recorded clock that ticked—even if the ticks were twenty-four seconds apart.

Why mechanize?

Water clocks can easily be calibrated to display the equal sixty-minute hours that make so much sense to us today. Yet in the heyday of the water clock, it was sundials, with their seasonally variable hours, that were the most reliable timepieces, and they dominated people's sense of time. All the sundials in the same town could be counted on to tell the same time on a sunny day when no two water clocks ever did. There was too much variability in their design, in the execution of their design, in the state of wear of the inlet aperture, and in the temperature of the water. Noticing how much water had run into a container remained a second-best way of telling the time—to be used as necessary at night, or on a cloudy day.

Burning the candle

Candle clocks were an alternative to water clocks for telling the time at night and in cloudy climates. Alfred the Great of England (849–899) is supposed to have marked the hours in this way. Candles were susceptible to drafts and they were expensive. This guaranteed that they were a little-used method. In China, evidence suggests that in some periods timed incense-burning was the dominant form of timekeeping, eclipsing the use of sundials.

Above: **A representation of how the candle timepiece invented by Alfred the Great would have worked, burning from one mark to the next.**

The time people wanted to know, anyway, was the time a sundial would tell, the time that people understood. So the wires in front of the planisphere in Athens' Tower of Winds showed the hours a sundial would have shown. Through ingenious feats of gearing, these were the sorts of hours announced by automated water-clock doors, bells, and whistles in wealthy Roman households and medieval Islamic cities.

Ironically, one of the chief motives for giving water clocks mechanized displays was to make them record sundial-based hours that we have abandoned in favor of equal hours that water clocks could have told unaided by wheels, gears, and pulleys. Another motive, of course, was to create impressive civic or private displays, and another was to hear the time. It was much harder to make a sundial trigger a pop, a whistle, or a bell.

Weight-driven bell-ringers

It was the wish for an audible time signal that led to the devices that are the direct ancestors of present-day clocks. After the 1270s, certain monasteries began using weight-driven, automatic bell-ringers to call monks to the divine offices at the prescribed times. They had no dials—what mattered was the ringing of the bell. The machines had to be constantly adjusted to keep in time with the changing length of day, as the Canonical hours, such as Lauds (before daybreak), Prime (at sunrise), Vespers (at sunset), and Compline (at nightfall) moved with the seasons.

Many languages named these machines after the time unit—the hour—they sounded. The French called them *horloges*, the same as sundials and water clocks. Similarly, the Germans called them *Uhren*. Other languages named them after the bells they rang. In Dutch (*klok*), Danish (*klokke*), Norwegian (*klokka*), and Swedish (*klocka)* the word for a bell and for a clock is one and the same. Medieval English borrowed "clock" as a foreign word. The direct source was probably northern European Church Latin, which in the 700s had borrowed the word *clocca* ("bell") from Germanic roots.

The verge and foliot escapement

Monastic bell-ringing machines and other late-medieval clocks worked by "escapement," which was a process of interrupting and then releasing the force of a weight suspended on a chain wound around a drum. As the chain unwound, the drum turned the machine's gears. To regulate the time it took the chain to unwind, the monks attached an adjustable escapement mechanism consisting of a "foliot" welded to a "verge."

The drum was slowed by the interrupted turning of a toothed "escape wheel" that could rotate only a small distance before one of its teeth caught on and had to push aside a flange on a vertical rod, the verge. Pushing aside this flange rotated the verge enough to put a second flange, set at a different angle, in the way of another of the escape wheel's teeth. The second flange was pushed aside in the opposite direction, so that the escape wheel pushed the verge first one way and then the other.

The foliot was a horizontal bar welded at right angles to the verge. As the verge part-rotated back and forth, so the foliot swung back and forth. The rate at which the foliot swung (and so the rate at which the escape wheel was allowed to push the verge back and forth) could be adjusted by positioning small balancing weights on the arms of the foliot. When the weights were moved outward along the foliot the clock ticked more slowly. When they were move in, it ticked faster.

Clocks in towns

In China, a technology like the one that culminated in Su Sung's Heavenly Clockwork could be snuffed out by a change of dynasty. So complete was the emperor's monopoly over the means of keeping track of time that the decision whether or not to develop this invention further was one entirely for court officials, on the basis of what would best serve the prestige of the reigning emperor. In the disorderly continent of Europe, no one person had an effective monopoly on ideas or inventions, or on any other source of power. Kings, bishops, barons, and emerging towns were in perpetual competition with each other for advantage, always ready to seize on any new way of bolstering their prestige.

Mechanical clocks were quickly and widely dispersed. Kings and bishops installed them in their palaces and towns put them in church towers, first in Italy (the church of Sant'Eustorgio in Milan had one in 1309), then all over Europe. Soon civic towers also had ticking bell-ringers. The appeal of automatic bell-ringing was the same for the towns as for the monks. In the largely illiterate pre-industrial world, bells were the prime channel of civic communication. Bells rang the hours, knelled for the dead, warned of enemy attack, called for firefighters, and celebrated the birth of a ruler's heir. To have some of this bell-ringing automated was a great convenience. It was also impressive and a mark of prestige.

Monasteries and abbeys quickly began adding features to marvel at in their clocks. So too did the towns. The first treatise on weight-driven clockmaking, by Richard of Wallingford (c.1292–1336), the Abbot of St. Albans, appeared in the middle of the fifteenth century. Wallingford was the son of a blacksmith and his treatise reveals that the clock he constructed for the abbey sometime after 1327 was mounted in a tower, and had a face with dials for the changing positions of the stars, sun, moon, and tides. Ostensibly a way of demonstrating the rationality of God's Creation, this tower display must also have enhanced the abbey's esteem.

By 1350, the city of Strasbourg had provided its cathedral with a magnificent clock that rang the hour while playing a tune. It also showed

the fifteenth century, rman craftsmen were king domestic clocks ich included foliot apements.

the date and current positions of heavenly bodies, as well as a procession of the Magi bowing before the Virgin Mary, followed by an enormous cock in wrought iron and copper that stuck out its tongue, crowed, and flapped its wings. This display was considerably embellished when the clock was rebuilt in 1574.

There were also private wonders. In 1381, the Milanese statesman Gian Galeazzo Visconti (1351–1402) purchased an astrarium (star indicator) from the clockmaker Giovanni de' Dondi (1318–89) who had spent sixteen years building it. A separate dial charted the position of each of the known planets. It had a twenty-four-hour clock dial, as well as one for sundial time in Padua, where the astrarium had been built. Its calendar included the dates of movable feasts such as Easter. It also gave the times of sunrise and sunset, and details of the solar and lunar cycles. It would attract admiring visitors to the Visconti castle in Pavia for more than one hundred years.

The rise of equal hours

With the arrival of clocks, the most immediate change in the way people thought was a new system for measuring the day: by twenty-four equal hours. Some public clocks were striking these hours by about 1330. If a town had many clocks and they followed different systems, the result could be confusing, and a royal decree was necessary in Paris in 1370 to make all the city's clocks chime twenty-four equal hours.

The pocket timepieces wealthy people carried about with them continued to be sundials that divided the daylight into twelve seasonally unequal hours. The Books of Hours that more and more people were now able to afford in order to guide their daily devotions referred to a third system—the Canonical hours still chimed in rural monasteries.

It was not because the new clocks were more accurate than sundials that equal hours were adopted. Just like water clocks, weight-driven timepieces had to be reset frequently. In spite of the adjustability of the balancing weights on the arms of the mechanism's foliot, the worst-made machines lost or gained up to an hour a day. The average was about fifteen minutes. The real reason for the shift to equal hours reflected the rise of commerce. For example, when textile manufacturers in Italy hired workers by the day, labor costs rose and output shrank in the short days of winter. In the 1330s, the industry began employing workers by the (sixty-minute) hour instead, and belfries were built to chime the hours so that workers did not have to rely on the rising sun for a time signal.

For town life, the Canonical hours rung in monasteries were not a priority. For the first time, neither were the seasonably variable hours shown by a sundial. Towns were growing wealthy enough to finance impressive civic features like clocktowers because of trade and manufacturing, and the traders and manufacturers who financed the clocks were people who valued time in a new way. They were attracted to time that could be accounted for consistently.

In 1335, the bell-tower of the Chapel of the Blessed Virgin in Milan was fitted with gears that caused hammers to strike the bell as many times as the number of the hour, from one to twenty-four. The first clockdial to show all twenty-four hours appeared in Italy in 1344. This system spread, though the rest of Europe experimented by dividing the day into four sequences of six hours or, more commonly, two sequences of twelve, starting one hour past midnight and one hour past the Canonical hour of Nones (noon).

Soon town-dwellers almost anywhere in Europe could tell simply by listening for the bells not only that it was the hour, but which hour it was. If they were literate they could often look up at a dial, read what hour had most recently sounded, and judge what portion of an hour had passed since. Countryfolk began remarking upon the curious way in which, in towns, people allowed themselves to be ruled by the clock.

Portable timepieces

In the four centuries from 1270 to 1670, little was done to improve the inner workings of weight-driven clocks. They continued to be hammered and riveted together by the same arts that blacksmiths had first used on monastery bell-ringers, and they continued to keep time very approximately. Specialized craftsmen traveled from one great city to another building public clocks, but most of their skill went into external displays. A more settled clockmaking profession also grew up over these centuries, however, and the demand it tried to meet was for a small, practical timepiece for the homes of the wealthy, or even for their pockets.

The Italian architect Filippo Brunelleschi (1377–1446) seems to have made a compact spring-driven clock in 1410. Many others were made by anonymous craftsmen financed by wealthy patrons, though a clockmaker's guild did not appear in Paris until 1544, in Geneva until 1601, and in London until 1627. The British Museum in London holds the oldest surviving spring-driven timepiece, a iron chamber clock in a gilt brass case, probably made in France in about 1460. It chimed the hours but had no face. The German clockmaker Peter Henlein of Nuremberg is credited with making the first watches, sometime in the first half of the 1500s. Spring-driven domestic timepieces were no more accurate than weight-driven clocks in towers, and their workings were not even enclosed against dust and moisture until early in the 1600s.

The direct social impact of spring-driven timepieces was much less than that of the weight-driven clocks, for very few people ever saw one. As late as the 1660s, the English diarist Samuel Pepys (1633–1703) records a life in which he kept time by churchbells, with occasional reference to a sundial. Although he was a government official (he would be secretary to the Admiralty by 1672), Pepys didn't own a watch in the 1660s, and he very seldom made appointments. Instead he circulated through London coffee houses and taverns, running into people he needed to see. In the main, the business to be done got done, but only when the opportunity arose.

A minority, however, were living a different sort of life. Among those wealthy few who in the 1600s could afford a clock on the mantelshelf or a watch in their purse, many were commercially minded, and valued time in a different way than a government official might. Before the 1300s, time had flowed evenly, on the analogy of sunlight streaming onto a sundial, or water flowing through a water clock. The Roman expression for wasting time was *aqua perdere* (to lose water). As clockwork became more common in European cities and fine houses, the commercial classes found themselves living in a time that ticked past. Even if seconds were not yet counted, the audible tick of their proud possession reminded them, as the moments passed, that they were spending something valuable. Not long after Samuel Pepys's diary entries for the 1660s, the word "punctual" in English took on the meaning that it has today.

Early compact timepieces had an impact on European society as well through the industrial skills and techniques pioneered by the craftsmen who made them. The precision machining they were forced to invent would underpin European mass production in the coming Industrial Revolution. Gears could be cut by hand for the great weight-driven clocks in churchtowers. For compact timepieces, though, tiny, accurately spaced gears, cleanly cut, were needed in a quantity that handwork could not supply. The first machine for cutting gears was made by Juanelo Torriano (1501–75), a clockmaker from Cremona in Italy.

Medieval screws were usually hand carved from wood. Churchtower clocks could be riveted together, but compact timepieces needed metal screws, and they needed them to be tiny. Lathes for cutting threads into screws appeared in clockmakers' workshops for the first time in about 1480. Cutting the grooves into a fusee was not automated until 1741.

The "fusee" makes springs keep time

To make clocks more compact, a coiled metal strip called a mainspring was substituted for weights as the driving force; but the problem with springs was that the power they released became weaker as they unwound. The solution was the "fusee," invented by military engineers in about 1400 to make it easier to draw a crossbow.

A fusee is a grooved, cone-shaped spool that increases the force transmitted by a drawn cord as it unwinds from the spool. In clockwork, it is fixed to the first gear wheel in the train of gears that turn inside the clock. The uncoiling spring pulls at a cord wound around the fusee. At first, when the spring is still fully wound, the cord unwinds from the thinnest end of the fusee, but eventually, when the spring is running down, it is unwinding from the thickest end. This evens out the force transferred through the cord from the spring to the gear.

Accurate time on land

Everyday clocks and watches of the 1600s were good enough for commercial purposes—even if pocket watches and public clocks had to be reset daily and were always at least a minute or two off. The drive for accuracy now came through the emerging needs of scientists. Governments and shipowners looked to science for a practical solution to the problem of longitude (see Chapter Seven), and clocks that were minutes out in their record of the difference in time between two points on the globe would produce calculations of longitude that were hundreds of miles in error. Myriad other calculations that scientists wanted to make—to predict the future positions of heavenly bodies, or predict when any recurring phenomena would reoccur—were severely limited if timed observations were not precise.

The great advance of the 1600s was to substitute a pendulum for the foliot that had regulated the pace at which a clock's driving force is interrupted and released. Galileo noticed, by watching swinging altar lamps at Mass, that unlike a foliot, a pendulum has a natural oscillation of its own. No matter how great the force of the swing, it takes the same time, depending solely on the pendulum's length. The French scientist Marin Mersenne (1588–1648) noticed this first, but it was Galileo (a year before he died in 1642) who first made a design for a clock in which the rate of turning of the gears would be precisely governed by the length of its swinging pendulum.

The first pendulum clocks were made in the Netherlands, starting in 1657, to a design by the Dutch astronomer Christiaan Huygens (1629–93). In about 1670, a much improved pendulum clock appeared. Instead of transmitting its rhythm through a verge, the top end of the pendulum had an "anchor" with a flange on either end that dipped rhythmically in and out of the escape wheel's teeth as the pendulum swung back and forth.

A commercial demand for pendulum-regulated domestic timepieces quickly developed. The average daily error of the household clock dropped dramatically from fifteen minutes to fifteen seconds. Most accurate were the weight-driven longcase ("grandfather") clocks, which could be set and left for several weeks. This new accuracy in clocks created a new demand for accuracy in sundials, which were expected to live up to new expectations. Clocks were still set to local time, which differed from one village to the next, and the arbiter of local time was the sundial. The difficulty in reading the shadows cast on most dials meant that clocks were typically set up to a minute wrong, even before they had a chance to lose or gain any seconds of their own.

Another problem was the variability in the speed with which the sun crosses the sky (a reflection of the fact that the earth's orbit around the sun is elliptical). Even after adjusting for seasonal differences in day lengths, differences in the speed of the sun's passage can result in errors of several minutes when converting to time in equal hours from the time shown on a sundial. By the late 1600s, clockmakers were supplying comprehensive tables to their customers that saved them from making the risky calculations themselves.

Minutes and seconds

With accurate, pendulum-regulated workings, it became practical to include a minute hand on clockfaces, starting in about the 1660s. The word "minute," little used in everyday speech before then, derives from the Latin *pars minuta prima* ("first division into the very small"), which describes the division of a degree (one three-hundred-and-sixtieth of a circle) into the sixty parts geometers call "minutes of arc." Astronomers divided the hour as well into sixty minutes in the 1200s.

The word "second" derives from the Latin *partes minutae secundae* ("second division into the very small"), namely the division of a minute of arc into sixty seconds of arc. In the 1500s, the term was adopted as well for one-sixtieth of a minute in time. By use of the pendulum, scientists began in the 1600s to time observations by one-second or two-second intervals. Second hands began to appear on clocks and watches in the 1700s.

Accurate time at sea

Pendulum-regulated clocks quickly proved incapable of keeping accurate time in vessels pitching about on the high seas. What was needed to solve the problem of longitude was a clock that did not rely in any way on gravity to make it work. Using a mainspring proved to be the answer in terms of providing a driving force. The problem that remained was to find an escapement regulator that was equally independent of gravity.

The answer was a balance spring attached to a balance wheel that oscillated back and forth around its own center of gravity. The English chemist and physicist Robert Hooke (1635–1703) was in dispute with Huygens over the claim to have invented this idea. Hooke's Law formulated the principle that makes the balance spring possible: "A spring when stretched resists with a force proportionate to its extension." But Huygens actually put the principle to work in 1674.

By the 1700s, spring-driven watches and clocks were greatly improved but still losing or gaining an average of a minute per day. They lost time in hot weather and gained time in cold. Another contributory factor to inaccuracy was friction in the escapement mechanism. The first clockmaker to solve these problems was the English carpenter turned clockmaker John Harrison (1693–1776). His breakthrough came

ristiaan Huygens
s an astronomer who
neered the use of
ndulum clocks.

through improvements to the escapement mechanism and by an arrangement of brass and steel rods that compensated for changes in temperature. After making the most accurate pendulum clock yet in 1728, he was attracted by the prize of £20,000 (an enormous sum in that day) that the British Parliament had been offering since 1714 for a solution to the problem of longitude.

To win the prize, the proposed method would have to prove itself on a trial voyage to the West Indies and back. The accuracy target set by the Parliament meant that any timepiece entered for the prize could lose or gain no more than two minutes in about six weeks of sailing. Harrison's first marine chronometer, tested on a voyage to Lisbon, was so promising that the Board of Longitude advanced him £500 when he said he would rather build an even better clock before entering properly for the prize. The Board waited impatiently while Harrison spent twenty-five years building three more chronometers, each one better than the one before. The one he finally sent for trial in 1762 was a large watch, about five inches in diameter. It lost only seconds on the journey to the West Indies, but Harrison had to struggle for ten years (submitting to a further trial, and receiving personal support from George III and members of Parliament) before the Board of Longitude finally granted him the bulk of the prize.

An age of synchrony

The method of finding longitude at sea by marine chronometer was not fully established until the 1820s, when the problems of mass-producing such fine timepieces had been overcome. The solutions drew more heavily on the designs of the French clockmaker Pierre Le Roy than on Harrison's.

By this time affordable, highly accurate spring-balance watches were among the possessions of very ordinary people, and in an increasingly synchronized world ordinary people found they needed them. Guards on British mail coaches were issued with watches in the 1780s, so that the coaches could run to a schedule. These watches were preset to gain or lose time, depending on whether the coach would be traveling east or west. In this way, the coach's time was constantly in synchrony with local time. In 1844, the first patent was granted for a clock that recorded the arrival and departure times of employees. French officers synchronized watches for the first time in the history of combat before their attack on Malakoff in 1855, during the Crimean War.

By the mid-1800s, railways in Britain were setting their schedules to Greenwich Mean Time, relayed by the Greenwich Observatory twice a day (at 10 A.M. and 1 P.M.) through the railways' telegraph system. This was the point when sundials became obsolete, because finding out what the local time was no longer mattered. People set their clocks by GMT and began rising and going to bed at the same time in the West of England as on the East Sussex coast.

In the U.S. and Canada, railway operators were soon steaming across a continent and demanding recognized time zones. This helped to bring the industrial nations of the world together in Washington, D.C. for the Prime Meridian Conference of 1884. It established GMT as a global standard, and carved the planet into the time zones we recognize today. For the first time, it became possible to know what time it was anywhere in the world.

The marine chronometer was obsolete before it had been in use a hundred years—the U.S. Naval Observatory began broadcasting daily time signals in 1905. Today, a Global Positioning System (GPS), including a network of orbiting satellites, can pinpoint to the second of arc the longitude and latitude of any object on earth. By the 1980s, it was possible to buy a radio-controlled wristwatch, a compact battery-powered timepiece in which the escapement was run by an oscillating quartz crystal and corrected by radio signal.

arrison's number
e chronometer, from
ound 1760.

Scientific time

Science has continued to lead the drive for accuracy. Because of fluctuations in the rotation of the earth, including a slowing of about one-and-a-half milliseconds a century, the earth's daily rotation was replaced as the scientific standard in 1955. The new standard was Ephemeris Time, based on the length of the year. This was replaced in 1967 by the caesium-ion atomic clock, which is independent of any astronomical observation and loses only one second in ten thousand years. Coordinated Universal Time, an averaging by the International Time Bureau in Paris of signals from eighty atomic clocks in twenty-four countries, is now the world standard time. Never easily contented, scientists in the early twenty-first century are looking forward to an "ion trap" timekeeper that will be accurate to one second in ten billion years, roughly the age of the universe itself.

9: journeys into deep time

Michael Roberts

Revising the age of the world

laid down as recently as 260 million years ago and as long as two billion years ago.

Some drill and bore
The solid earth, and from the strata there
Extract a register, by which we learn
That he who made it, and reveal'd its date
To Moses, was mistaken in its age William Cowper

One of the most dramatic sights in the world is the Grand Canyon. Nowhere demonstrates more vividly the abyss of time. To stand at the rim and glimpse the Colorado flowing 5,000 feet (1,500 meters) below is to see a snapshot of the passage of time. The Precambrian rocks at the bottom are nearly two billion years old. Above that and succeeding each other like layers in a cake are strata through most of the geological ages from the Cambrian (550 million years) to the Permian (283 million years). The rocks record a time duration that is beyond comprehension. The Grand Canyon is a monument to deep time.

Deeper into time

Four hundred years ago the earth was thought to be just a few thousand years old. With the gradual development of scientific understanding, man began to comprehend the abyss of time. According to Archbishop Ussher in 1656, the earth was created on October 22, 4004 B.C.—a date that has come to signify the traditional date of Creation according to the Church. Geologists and Darwin had to battle against this idea.

The discovery of deep time is not simply one of Judaeo-Christian time giving way to scientific time after a long and bitter struggle. The picture is more complex, as religious ideas have both hindered and contributed to the development of our concept of time. The scientific understanding of time, whether geological or astronomical, developed in Western Europe from about 1650, in societies which had a dominant, but declining, Christian culture. As a result, Christian thought had an important influence on the understanding of time.

The linear understanding of time by Christians in the seventeenth century

Creation	Man	Abraham	Moses	David	Exile	Jesus	Second Coming
	4004	2000	1500	1000	590	4 B.C.	?

In an apparent paradox, concepts of circular or cyclical time gave way to linear time. Linear time has roots in the biblical tradition with an initial act of Creation, followed by acts of God in history, the Flood, Abraham, Moses, David and the Monarchy, the Exile, and finally Jesus Christ. Cyclical time springs from the Greek and Eastern traditions and surfaces in the skepticism of the eighteenth century where the eternal nature of the physical universe was emphasized against the finite, linear Christian view. Christians accepted the linearity of time long before it became scientifically incontrovertible.

The story of the scientific examination of the age of the earth and the universe goes back 350 years to 1650. Before that there are, at times profound, philosophical discussions on time, but these are not scientific and give no sense of duration. Before looking at these it is essential to consider Christian understandings, which were prevalent at the start of the scientific revolution in the seventeenth century.

The Book of Genesis

The stories of Genesis are well-known in the western world. However, few people reflect on what Genesis meant to the writer(s) or when it was written. The answer to the second question: most biblical scholars reckon that the book was compiled from earlier sources in Mesopotamia in about 500 B.C.; a minority believe it was compiled about 1000 B.C. Many Evangelicals hold that Moses wrote Genesis and the Pentateuch in about 1250 B.C. Some claim that Genesis is derivative because it has parallels with other near eastern creation myths, such as *The Epic of Gilgamesh*.

Exactly how the Hebrews understood Genesis is harder to answer. Some scholars argued that the universe was a three-decker one and the representation of time is literal. If the Hebrews were as informed on astronomy as their neighbors in Egypt and Mesopotamia, they would not have held such crude ideas. As the Old Testament writers borrowed ideas

·eation stories: Detail of .e *Creation of Adam*, · Michelangelo, from e Sistine Chapel.

from both civilizations—for example, Proverbs 22 is borrowed from the Teaching of Amenope—they probably knew their astronomy as well. However, few studies focus on what the ancients believed about time and this may be an unanswerable question. Without entering the controversy about whether the Old Testament is historical, it is safe to say that the Hebrews were not as concerned as later societies about chronology.

The early Church

The early Christians, on the other hand, were interested in throwing light on biblical chronology. Until A.D. 400, the vast majority believed that the earth would last only 6,000 years and had existed for about 5,500 years when Christ was born. They took all biblical chronologies, especially those in Genesis 5 and 11, literally. "Chiliasm"—the popular belief that the earth would last six days of millennia (from Psalm 90 vs 4 and 2, Peter 3 vs 8)—was proclaimed, rather than reasoned, in the Epistle of Barnabas (c. A.D. 130), "Therefore, my children, in six days—six thousand years, that is—there is going to be the end of everything." This concern with the end of the world may explain the early Christians' great interest in chronology.

An early example of this interest is *Ad Autolycum* by Theophilus of Antioch, of whom little is known except that he became Bishop of Antioch in A.D.169, and wrote this volume after the death of Marcus Aurelius in A.D. 180. A Greek, he was strongly influenced by Jewish Christians. The work finishes with a chronology from Creation to the death of Marcus Aurelius, a duration of 5,695 years, suggesting that Creation occurred in 5515 B.C. The chronologies are calculated from biblical data, and are not far off Ussher's compilations and today's estimates from Abraham to the Exile.

Theophilus was highly literalistic, while others, like Augustine, interpreted the days of Genesis allegorically; few thought the earth to be more than a few thousand years old.

Comparison of estimates of biblical chronology by Theophilus (180), Ussher (1650), and in the Good News Bible (1976). The date for Noah is taken from Pitman and Ryan, Noah's Flood (1999).

Event	Theophilus	Ussher	Good News Bible
Exile	561	593	587
Death of David	1079	1014	970
Moses	1577	1491	c.1250
Abraham	2262	1921	c.1900
Noah	3273	2347	5600
Creation	5515	4004	?

A version of the creation of the world, taken from the cupola in the atrium of San Marco church in Venice, Italy.

The Renaissance

The Renaissance was a time of broadening horizons and exploration. Columbus discovered the New World; Copernicus rejected the Ptolemaic system and proposed heliocentricity as the best explanation of the relation between the sun and planets. There was a revival in the study of ancient texts, classical and biblical, which resulted in the Reformation. In this flowering of exploration, scholarship, and literature there was a sense of the unity of knowledge.

It also marked the dawn of an historical consciousness, but concepts were scarce and the Scriptures were among the few texts that went back to the earliest history. So attempts at the history of the world involved the fusing of biblical and classical writings—for example, Sir Walter Raleigh's *History of the World*, published in 1614. Raleigh placed Creation at about 4000 B.C.—the same date proposed by the sixteenth-century Protestant reformer Martin Luther (1483–1546), the Roman Catholic Cardinal Bellarmine (1542–1621), and the devisor of the map projection, Mercator (1512–94). A century earlier, Columbus was more generous with 5443 B.C. The date range 4000–5000 B.C. was widely accepted as the approximate time of the origin of the earth. Most Protestant and Roman Catholic theologians concurred on about 4000 B.C. and the Geneva Reformer John Calvin (1509–64) typically reckoned "the present world is drawing to a close before it has completed its six thousandth year."

Renaissance dates of Creation

Columbus 1480	5343 B.C.
Mercator	4000 B.C.
Luther	4000 B.C.
Bellarmine (RC)	4000 B.C.
Raleigh	4000 B.C.
Scaliger c.1610	3950 B.C. (Sun Oct 25)
Ussher c.1656	4004 B.C. (6 P.M. Sat Oct 23)

Chiliasm

As the Reformation progressed, a revamped form of Chiliasm developed. In the early 1600s, the Dutch Protestant theologian Josef Scaliger (1540–1609) put the date of Creation at October 25, 3950 B.C. (The fall was a favored time for Creation, as the fruits would provide sustenance for the winter.) The best-known Chiliast was Archbishop James Ussher of Armagh. His *Annales Veteris Testamenti* (1650) was a solid piece of chronological scholarship in which he argued on historical grounds that Jesus was born in 4 B.C. But he is remembered for his date of Creation—4004 B.C. He did not conclude this from arithmetic applied to dates of patriarchs and Old Testament figures; to Ussher there were six Chiliastic days of 1,000 years followed by the seventh day of the Millennium.

Ussher's Chiliastic days

The Seventh Chiliastic Day
The Millennium—began 1996?
1996

The Second Chiliastic Day of the Christian dispensation
A.D. 996

The First Chiliastic Day of the Christian dispensation
4 B.C.

4 B.C. The Birth of Christ

The Fourth Chiliastic Day of the pre-Christian dispensation
A.D. 1004

The Third Chiliastic Day of the pre-Christian dispensation
A.D. 2004

The Second Chiliastic Day of the pre-Christian dispensation
A.D. 3004

The First Chiliastic Day of the pre-Christian dispensation
A.D. 4004, Creation: 6 P.M. Sat October 22, twelve-hour chaos

There were four Chiliastic days before Christ, and so Creation took place in 4004 B.C., on the night before October 23. Adam was created on October 23. This date seems quaint, but the rest of Ussher's chronology was very sound for the seventeenth century. His chronological calculations for the rest of the Old Testament are close to modern estimates, and had they not been inserted in many English bibles from 1704, he would probably have been forgotten by all but historians, who valued his careful work.

Theories of the Earth, 1660–1710

The Royal Society of London, founded in 1660, epitomized the flowering of science both in Britain and on the continent. The work of Robert Boyle (1627–91), Isaac Newton (1643–1727), and others in physics and chemistry needs no introduction. Less well-known is the natural history of John Ray (1627–1705), Edward Lhwyd (1660–1709), and others. The period saw the beginnings of a scientific study of the earth; findings were published in turgid volumes known as *Theories of the Earth*. These seem at first to be literal readings of Genesis with a semi-scientific gloss. A closer look shows them to be more profound as they meld together the Bible and the classics in a style reminiscent of medieval "book" learning, with the citing of endless authorities and scientific insight. They shared the outlook of most theologians—except Ussher—and literary writers such as Thomas Traherne (1637–74) and Alexander Pope (1688–1744). Instead of taking the Creation story to teach that Creation took place in six short days, writers followed an interpretation dating back to the early Church Fathers in claiming from Genesis (Gen 1:1) that God first created Chaos (without form and void) and after an interval recreated it in six days. The duration of Chaos was undefined.

With Ussher it was twelve hours, but for most it was long and unspecified. Some seventeenth-century writers—notably Thomas Burnet (c. 1635–1715), Edmond Halley (1656–1742), and William Whiston (1667–1752)—reckoned the days to be longer than twenty-four hours. Halley attempted to calculate the age of the earth from the salinity of the sea, but reached no firm conclusions other than that it was tens of thousands of years old. Contemporary theologians had similar theories; Bishop Simon Patrick thought that God first created Chaos and then later reordered it in six days. He said of the duration of Chaos, "It might be…a great while."

Few concurred with Ussher's date of 4004 B.C. for the initial Creation, although most accepted that humanity first appeared in about 4000 B.C., so the rest of Ussher's chronology was generally accepted. The extension of time by the "Theorists" and theologians was minute in the context of billions of years of geological time, but was—as Stephen Gould wrote of Whiston's argument that the day of Genesis One was a year long—undoubtedly "a big step in the right direction." In Britain, the way was open for a longer timescale.

lexander Pope was ne of those writers ho combined the iblical teachings ith a more scientific pproach.

Chaos, Creation, and Deluge in the seventeenth century

FINAL CONFLAGRATION		c. A.D. 2000
		4 B.C. Birth of Christ
Deluge (Genesis 6–8) of great geological import, 2400 B.C. General consensus (except Stillingfleet, Lhwyd)		
Fall (Genesis 3) Caterpillars and other noxious insects created—Calvin Earth tipped on axis—Milton		
Mankind created in		**c.4004 B.C.**
Day six	Land creatures and man	
Day five	Creatures of water and air	
Day four	Lights	
Day three	Fertile earth (Whiston: day = year)	
Day two	Sea and sky	
Day one	Light and dark	
		Often 4004 B.C.
Chaos *tohu bohu* Hesiod *First of all there was a chaos.* Also Ovid etc.		
Duration of Chaos	Ussher 1656	Twelve hours
Milton	Short?	
Ray	Vague	
Whiston	Long time	
Thousands of years ago		
In the beginning God		

Fossils and geology

It was not until the late seventeenth century that fossils were recognized as imprints of dead creatures rather than objects formed as a "sport of nature." Only then could they be used to demonstrate former life; it was a century before fossil classification was used to put rock strata into historical order. Possibly the first person to use the succession of fossils to explain evolution was Charles Darwin (1809–82), in a notebook in 1838, shortly before he discovered natural selection. In the 1690s, there

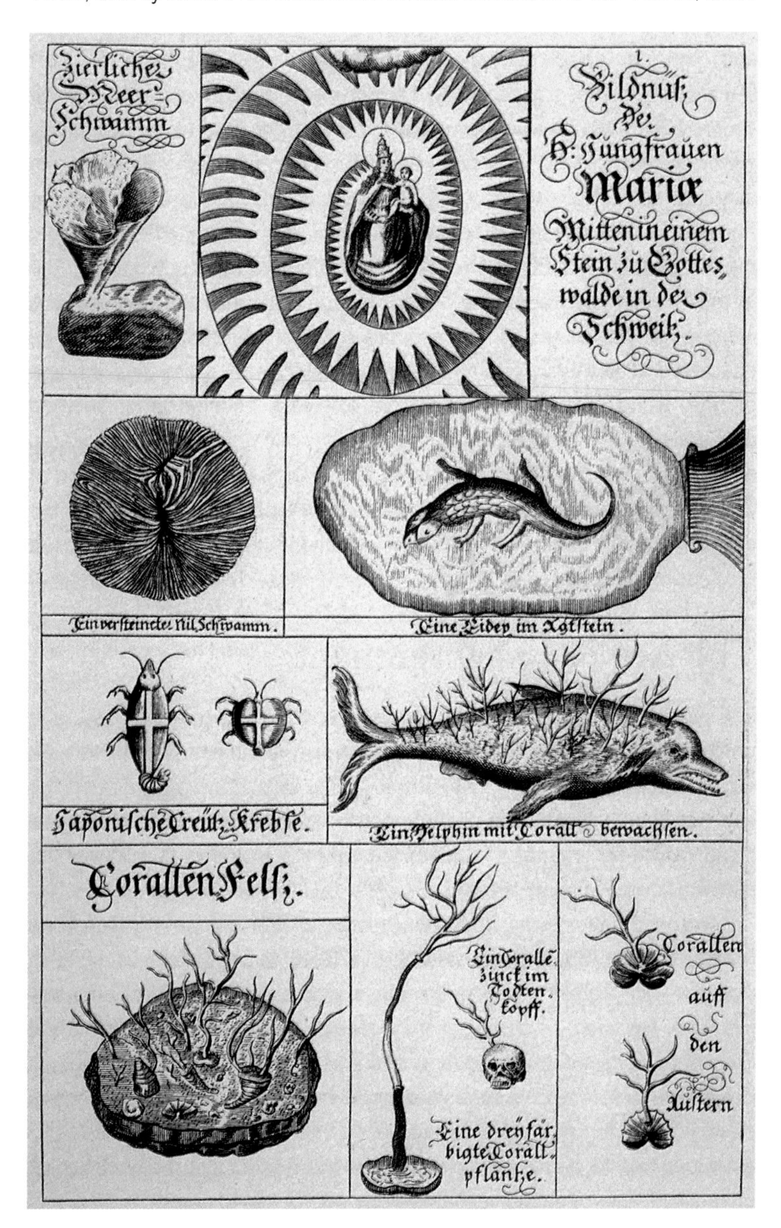

A set of illustrations of seventeenth-century fossils.

Creation. There were two creations: the creation of the basic materials—Chaos—and the creation of the universe with those materials, all of which God accomplished "in one hundred and forty-four hours," as in Genesis. Although the re-Creation took 144 hours, there is no indication how long the Chaos lasted. Most other religious writers held similar views; only a minority espoused a young earth. At the end of the eighteenth century, Joseph Haydn's oratorio *The Creation* was written with an orchestral introduction on *The Chaos* followed by the aria "And a new created world sprung up at God's command." The libretto of *The Creation* dates from mid-eighteenth-century England. An unknown poet took Milton's ideas in *Paradise Lost* and wrote it for Handel. In 1792, Haydn obtained a copy while in England and put it to music on returning to Austria.

Many poets incorporated Chaos when writing about Creation—an example of this is the poet Phillis Wheatley's (d.1784) *Thoughts on the Works of Providence*,

> *That called creation from eternal night.*
> *"Let there be light," He said: and from his profound*
> *Old Chaos heard*

Wheatley was an African-born slave adopted by John Wheatley of Boston. The Wheatleys—slave-owners and slave—moved in Evangelical circles and are more usually referred to in the context of abolitionism; but this excerpt indicates how widely held the concept of Chaos—and therefore the duration of time—was.

Hutchinsonian literalism

The views of the clerical scientist John Hutchinson (1674–1737) and his disciple Alexander Catcott (1725–79) were quite different from those held by Newton. Both lay great store by Genesis and sought to correct the "errors" of Newtonianism. In 1748, Hutchinson wrote *Moses' Principia* in direct opposition to Newton. Far less is made of the Chaos than in the *Theories* and Hutchinson seems to have believed that the Chaos or *tohu va bohu* was of no significant duration. In 1868, Catcott wrote his *Treatise on the Deluge* which implied that Chaos lasted only a short time. Hutchinsonian ideas endured until the early nineteenth century; the last Hutchinsonian scientist seems to have been the entomologist William Kirby (1759–1850), who argued for a six-day Creation in his *Bridgewater Treatise*. It would be fair to see Hutchinsonianism as a biblicist reaction to the prevalent Newtonianism.

For the first three-quarters of the eighteenth century, there was no consensus on the age of the earth. The case of Phillis Wheatley should caution against the assumption that it was commonly thought to be a mere 6,000 years old, as only the literate have left any evidence. A minority took the Bible literally, adhering to an Ussher chronology, but most Christians, whether Evangelical or not, stretched this belief to hold that the Chaos lasted an indefinite period of time with humanity limited to 6,000 years. It is difficult to decide whether the lines of William Cowper, an Evangelical poet who also wrote a poem of appreciation to Erasmus Darwin, reflect a concern for geology,

Some drill and bore
The solid earth, and from the strata there
Extract a register, by which we learn
That he who made it, and reveal'd its date
To Moses, was mistaken in its age.

William Cowper, *"The Task"*

Eighteenth-century views of the age of the earth

1. Scripturally derived figure of 6,000 years since restitution of Chaos in six days. Preceded by the existence of the created Chaos of undefined duration. Thus Creation is much earlier than 4004 B.C.. Mankind only 6,000 years old.

   The most widespread view of educated Christians. Most theologians. Many poets, Haydn's creation. Lip service from "skeptics"–for example, Erasmus Darwin.

2. Scripturally derived figure of 6,000 years. Most commonly Creation in 4004 B.C..

   The popular view of Christians. Held by some conservatives, but relatively few, an example being Hutchinson.

3. Scriptural chronology preceded by geological periods with figurative long days. Mankind only 6,000 years old.
   A less commonly held view of educated Christians.

4. Those geological periods and extended history of mankind not limited by Genesis.
   Essentially held by Deists and similar people.

5. An eternal earth.
   Deists.

James Hutton was one of a few scientists who first looked deeper into the age of the earth.

The discovery of deep time

The vastness of time was little understood until the end of the eighteenth century. Although recognition for the discovery of deep time is often assigned to James Hutton (1726–97), the Scottish physician and scientist, the "discovery" was also made by several scientists in Europe in the last two decades of the century. These were the Genevan polymath and mountaineer Henri de Saussure (1740–99) in the Alps near Chamonix in 1778, the German mineralogist Gottlieb Werner (1749–1817), the Parisian palaeontologists Georges Cuvier (1769–1832) and Alexander Brogniart, and in England, the canal engineer William Smith (1769–1839), working near Bath. However, no individual should be given all the credit.

By 1800 the earth was known to be millions of years old. Bound up with this discovery was the development of the use of fossils to determine stratigraphy and the historical order of rocks. Before long the succession of life—and extinction—was apparent. The stratigraphic column was slowly worked out; it was the prime task of geologists until mid-century. The sequence of Cambrian, Ordovician, and other eras of strata appeared.

Yet no precise figure could be given for the age of the earth. De Saussure thought it was very old, his compatriot Jean André (J.A.) de Luc (1727–1817) thought it was tens of thousands of years old—yet, in the 1780s, Abbé Soulavie was denounced for impiety by fellow Abbé Barruel for allegedly estimating the earth's age to be 356,913,770 years. By 1820, an eccentric British clerical-geologist named William Buckland (1784–1856) was reckoning "millions upon millions" of years.

The Church made no concerted attack on these geological findings; most educated Christians accepted them happily, which was not surprising as many were clergy. John Playfair (1748–1819) of Edinburgh and Joseph Townsend (1739–1816) of Bath were prominent figures at the end of the eighteenth century, who publicized the work of Hutton and Smith respectively. Those churchmen who opposed the discoveries made by geologists were always in a small minority.

Christian accommodation

At the beginning of the nineteenth century, many Christian, or nominally Christian, writers modified the consensus of the Theorists. The sequence based on Genesis 1–11 (the initial creation of Chaos, the reordering of Creation in six days with man being created in about 4000 B.C., and then the Deluge) evolved into a vastly extended Chaos to allow for the vast time of geology and a multiplication of Deluges.

Theologians quietly slipped geology into the Chaos. The first of these seems to have been Thomas Chalmers (1780–1847) at St. Andrews in the winter of 1802. In 1816, the future Archbishop of Canterbury, John Bird Sumner (1780–1862), published similar ideas in *A Treatise on the Records of Creation.* Both Chalmers and Sumner were Evangelicals and of an intellectual bent. This harmonization of geology and Genesis was widely accepted—by Buckland, for example—and staved off any major conflict, but between 1820 and 1850 a minority tried to dismiss geology and insist on a six-day Creation. Their strongest opponents were the clerical geologists and their supporters.

Many Christians harmonized geology and Genesis in the early nineteenth century

Deluge	By 1820s—only a local flood The last thirty feet of Diluvium Caused by Ice Age—Buckland (1841)
FALL (Genesis 3)	Effects only on man, not the rest of the natural world

Mankind in c. 4000 B.C., being pushed back after 1820—for example, Herschel 1836

Day six	Land creatures and man
Day five	Creatures of water and air
Day four	Lights
Day three	Fertile earth
Day two	Sea and sky
Day one	Light and dark

Estimated 6,000 years ago—for example, by Chalmers 1802–10—and then extended

Chaos *tohu bohu*	"without form and void" (Genesis 1 vs 2). All geological strata deposited in this period

Tertiary
Chalk
Jurassic
New Red
Carboniferous
Old Red
Killas/Silurian/Cambrian
Many millions of years—an indefinite period

In the beginning God

From about 1810, English geologists were concerned mainly with determining the stratigraphic order of rocks before delving into philosophical questions. Most geologists accepted some kind of multiple Catastrophism, with Noah's Deluge as the last of these; they were known as Diluvialists. In the 1820s, some geologists (Fluvialists), notably Charles Lyell (1797–1875), rejected Catastrophism and suggested a more gradual Uniformitarianism.

Meetings of the Geological Society of London often became fiery debates between the Fluvialists, led by Lyell, and the Diluvialists, led by the Rev W.D. Conybeare (1787–1857). The port flowed freely at the associated dinner parties. These debates are often presented as if it were Lyell who introduced the notion of a great age; in fact, both Lyell and the "Conybeare Sect" (as Lyell called his friends), accepted vast geological ages.

Uniformitarianism led to the Deluge no longer being seen as geologically significant or as the last of many Catastrophes, but many geologists were not entirely convinced by the theory. Lyell scarcely affected opinion on the age of the earth.

Geology comes of age

For the first half of the nineteenth century there was an almost unanimous conviction that the earth was extremely old, but still no consensus of opinion about how old. Geologists were unraveling layer upon layer, strata upon strata going from the recent Drift rocks down through the Mesozoic to the Cambrian and below. The timescale involved was immense and seemingly immeasurable. Critics of geology picked up on the sweeping statements of geologists, who spoke of millions of years as if they were days.

Buckland asserted that the earth was "millions upon millions" of years old and his friend Conybeare that it was "quadrillions." Darwin said of his geology tutor Adam Sedgwick (1785–1873), "What a capital hand is Sedgwick for drawing large cheques upon the Bank of Time!" Beyond affirmations of great age, Lyell gave no figures. This consensus of millions upon millions of years contrasts with de Luc's figure of tens of thousands in the final decade of the eighteenth century. After the mid-1820s, even the most devoutly religious of geologists would never suggest such a low figure for the age of the world. The earth was millions of years old, but no one knew how many millions.

A diamond-backed rattlesnake lies asleep a bedding plane in Cambrian strata in th Grand Canyon, just below Indian Gardens Note a succession of t shales within sandstor

Reaction to deep time

Throughout the half-century when deep time was developed, a minority challenged it. In the 1790s, Richard Kirwan (1733–1812) famously opposed Hutton's vision of an eternal earth. In 1802, the same year that the French naturalist Jean Baptiste (J.B.) de Lamarck (1744–1829) developed his own theory of evolution, François-René de Chateaubriand (1768–1848) of Combourg Château in Brittany published the *Génie du Christianisme*, a Catholic literary *tour-de-force* reacting against the French Revolution. He rejected Buffon's long timescale, commenting, "*Dieu a dû creér, et sans doute créér le monde avec toutes les marques de vétusté*." This can be translated, "God had to create, and doubtless created the world with all the marks of antiquity and decay"—that is, the world may *appear* ancient, but it is actually a recent creation. This argument was later taken up by Gosse.

As biblicist Evangelicalism took root in Regency Britain, opposition to geological ages mounted, even dismissing such Evangelical geologists as Sedgwick as infidels. The outcry did not last, and by 1850 most educated people, whether Christian or not, had accepted the deep time of geologists. For decades, geology had been the most popular science. Clergy did sterling fieldwork in their spare time, geology books were available in mechanics institutes, and geology lectures were given in most towns and often in churches.

By 1860, scarcely any clergy or educated people rejected geological time. Almost the only person to do so was the naturalist Phillip Gosse (1810–88), whose life is portrayed in semi-fictional form by his son in *Father and Son*. Gosse was a man of extreme religious views; he was not a member of any mainline church. In 1857, he wrote *Omphalos* (Greek for "navel") to show that, although the earth appears old, Creation took place in an instant some 6,000 years ago. Thus Adam was created *de novo* (complete with a navel). The book was a lead balloon—the parson-naturalist Charles Kingsley (1819–75) said that it made God a liar.

If Gosse can be ignored as an extremist, Canon F. Maupied—a leading Roman Catholic theologian at the Sorbonne—cannot. In the 1840s, Maupied was teaching a literal Creation and a rejection of geology during his lectures, something that had never been done at Oxford or Cambridge. After 1850, apart from a few Evangelicals and Seventh Day Adventists, hardly anyone argued against deep time until 1961, when the Creationist movement took off.

The motivation behind this reaction to deep time lay in both theology and politics—reform in Britain and revolution in France.

Deep time and *The Origin of the Species*

When Darwin published *The Origin of the Species* in 1859, the vast age of the earth was an idea as established as heliocentricity, but no firm figure could yet be put to its age. To many physicists, geologists were far too cavalier in their guesses at geological time. The Reverend Samuel Haughton (1821–97), geology professor at Dublin and an ardent opponent of natural selection, suggested that 1,526 million years had passed since the beginning of the Cambrian—three times the present figure. Darwin was equally generous with time and suggested that some three hundred million years had passed since the early Cretaceous. He argued this on sound geological reasons, but also to allow sufficient time for evolution.

Although much is made of the (limited) religious objections to Darwin, the strongest opposition came from physicists such as William Thomson (later Lord Kelvin; 1824–1907), Peter Guthrie (P.G.) Tait (1831–1901), and James Joule (1818–89). Theirs was a chauvinism of "exact scientists" about the imprecision and lack of quantitative methods of geologists and biologists. Kelvin struck at the heart of both with his estimates of the age of the earth. From 1855, Kelvin reckoned it to be no more than one hundred million years—a minute fraction of the time Darwin needed for evolution to occur through natural selection. Most biologists and geologists succumbed to the authority of physics. In later editions of *The Origin of the Species,* Darwin removed all references to his high estimates, and so for the rest of the nineteenth century Darwinism as such was eclipsed. Most evolutionists adopted a non-Darwinian theory of evolution.

In the 1880s, Kelvin reduced his estimates to about twenty-four million years, making the process of evolution almost impossible unless an ineffable outside agent was controlling it. Kelvin had virtually bludgeoned geologists into accepting limits on the age of the earth and for half a century from 1860, few geologists dared to suggest a figure of more than one hundred million years. In 1860, John Phillips (1800–74), nephew of William Smith and geology professor at Oxford, suggested ninety-six million years. He estimated that the rate of deposition at the time was one foot (a third of a meter) in 1,332 years. As the estimate of the thickness of fossiliferous strata was 72,000 feet, that gave an approximate figure of ninety-six million years. This gave credence to Kelvin's 1868 estimate of one hundred million years. Huxley was less happy but unable to assail Kelvin's calculations. Few geologists were able to challenge Kelvin, because they had no good method of quantifying geological time and adopted guesstimates based on supposed rates of deposition.

artoon in the satirical publication, *Punch,* the nineteenth century. hows the development life from "chaos" ough all the different ges of man, from the rm to the supposed hest point—Charles rwin himself.

The thickness of strata in the various periods at least gave a good indication of the *relative* length of the periods. In 1893, three years before the discovery of radioactivity, William McGee took Precambrian sediments into consideration. Playing with the arithmetic, he allowed the age of the earth to be between ten million years and five thousand billion years and favored six billion, not far off the 4.54 billion years of today.

Despite this great disparity of estimates, the one agreement was that the age of the earth was to be measured in millions of years. This decision was shared by most Christians, including the Evangelicals, whose ideas of time were published in 1910 in the booklets titled *The Fundamentals.* Writers such as James Orr, a Scottish theologian, and George Wright, an American clergyman and geologist, used this limited time to show how evolution had to be divinely guided rather than occurring through chance and natural selection. But by then time was getting deeper still.

Deep time goes deeper

While Kelvin was shrinking the age of the earth, in 1896 the French physicist Henri Becquerel (1852–1908) discovered radioactivity, since uranium compounds emitted energy similar to x-rays. Radioactivity had two major implications for the age of the earth. The first was that radioactive decay created immense energy, negating Kelvin's arguments for a cooling earth. The second was that radioactive elements could be used to measure time as they disintegrated at a fixed rate—known as a half life. Kelvin went to the grave without accepting the implications of radioactivity, which destroyed all his arguments and soon supplied ways of estimating the age of rocks and of the earth, giving vastly greater figures. In 1905, the English physicist John William Strutt, later Lord Rayleigh (1842–1919), showed that a mineral containing radium was two billion years old because of its helium content. In the same year, Bertrand Boltwood suggested that lead may be the end product of the decay of uranium, and calculated the ages of forty-three minerals from four hundred to two thousand two hundred million years old. The radiometric dating game had begun.

Arthur Holmes and the age of the earth

For the next fifty years the most innovative geologist in the dating game was Arthur Holmes (1890–1964). He wrote articles on geological time and several editions of a short but profound book *The Age of the Earth* in 1913, 1927, and 1937. In 1913, he based his work on three uranium-lead results from the Palaeozoic. Combining this with the thickness of sediments, he estimated the base of the Cambrian to be six hundred million years, remarkably close to present figures of five hundred and fifty million years. His work shows outstanding geological insight. As time wore on, the number of age determinations multiplied and are almost infinite. A study of Holmes's work over half a century (as carried out by Cherry Lewis) shows how a scientific theory can gradually be supported by strong experimental data.

Initially Holmes reckoned that the earth was younger than two billion years old (actually 1.6 billion in 1913 and 1927, and 1.75 billion in 1937). From 1946, the figure was seen to be nearer 4.6 billion, with the Cambrian commencing in about 550–590 million years; the lower estimate is accepted today. Despite many refinements and an explosion of methods and age determinations, this has remained the same for half a century.

There are three basic methods of determining the age of the earth. The first is to date the oldest rocks, which gives a minimum age. The ages of samples of ancient rock called "gneisses," from Greenland, first "dated" by the Oxford geologist Stephen Moorbath and others in the early 1970s, have not yet been bettered. The five methods used give an average of 3.65 billion years. Nearby, there are some other whole rocks which date back to 3.8 billion years. But in the last ten years, tiny fragments of crystal, called detrial zircons, have given ages of up to 4.4 billion years, indicating that the grains may have been formed at that time, yet were deposited by water about 3 billion years ago. If that is so, it would mean that the earth had cooled to form a crust much earlier than previously thought, about two hundred million years or so from the formation of the planet. The second way of finding the age of the earth is to examine meteorites, which date between 4.5 and 4.7 billion years. The third is theoretical and involves determining "model lead ages" from the decay of uranium into lead for the earth, the moon, and meteorites. This method was developed independently by Holmes and Houtermans in 1946.

Astronomy becomes cosmology

In the early modern period up to 1800, astronomers were more interested in matters of space than of time. First of all, geocentricity was rejected and then elliptical—rather than circular—orbits were accepted. It was possible to calculate the distance of stars by geometrical means. As estimates had been made of the speed of light since the seventeenth century, by 1800 the distance of some stars indicated that they were two million light years away—indicating a vast age of the universe in harmony with geological estimates. Beyond this there was little sense of time. The first estimates on the age of the universe were made in the 1920s, nearly two decades after radiometric age determinations showed the earth to be several billions of years old.

These initial estimates about the universe were possible thanks to the discovery of the redshift-distance relation—Hubble's Law—in the 1920s. Edwin Hubble (1889–1953) worked at the Mount Wilson Observatory above Los Angeles with Milton Humasson, an astronomical photographer. Hubble's Law pointed to the universe forming at a particular time. From this point, the age of the universe became an important and potentially answerable question. Over the next seventy years it became a perennial issue in which the "age" depended on the value of Hubble's Constant (H). Ironically, this time the astrophysicists had to see that their results concurred with the geologists' age of the earth of several billion years, in marked contrast to the time when the physicists in the person of Kelvin overruled the geologists.

The Belgian astronomer-priest Father G. Lemaitre (1894–1966) began to talk of a beginning in time for the universe, whereas in 1932 Einstein had avoided this question. In Rome in 1952, W.H.W. Baade (1893–1960), a German émigré living in the U.S., argued that the universe was between 1.8 and 3.6 billion years old—somewhat less than, but to the same order as, estimates for the age of the earth. All hinged on the value of H. Over the years, then, estimates for H have varied from forty to 525, with suggested ages up to twenty billion years. In 1997, estimates lay between ten and thirteen billion years, with a best value of 11.5 billion. And that figuring clearly concurs with the geologists.

However, a philosophical or theological question remains. What was there before the birth of this universe? It was Father Lemaitre who implicitly raised this question—which also has theological implications.

The duration of *Homo sapiens*

Until the nineteenth century, few doubted that humans had existed for only six thousand years, or a similarly short time. In the absence of archaeology and with limited historical research, this is not a surprising assumption. Even when geologists were pushing back the age of the earth, six thousand years for the time span of humanity was still accepted. This is often attributed to the influence of the Church's teaching about the age of the earth—for example, the claim about human remains discovered at Paviland Cave in South Wales. In the 1820s, William Buckland deduced that these human remains, which he christened the

The Hubble space telescope.

Scarlet Lady of Paviland, dated from Roman times; but later work showed them to be twenty thousand years old. This was partly because Buckland believed that the existence of humanity was relatively recent, although he was liberal on matters relating to geological time.

During the early nineteenth century, evidence slowly came to light, so that by mid-century many accepted that human history was older. By 1847, geologists Edward Vivian (c. 1820–93) and William Pengelly (1812-94) were suggesting that the human remains at Kent's Cavern in Devon were far older than the Deluge, but their ideas were not acceptable to the Geological Society. In 1831, Charles Lyell held that humankind was recent (namely, six thousand years old), but in *The Antiquity of Man* (1864) he extended this age to one hundred thousand years. It remained a controversial point for many years and even in 1912 George Frederick (G.F.) Wright (1835–1921) was arguing that "the antiquity of man…need not be more than fifteen thousand years old." That figure was almost immediately untenable owing to radiometric age-dating. Dating the appearance of *Homo sapiens* is still a controversial subject and is now believed to have taken place one hundred thousand to one hundred and fifty thousand years ago.

The cosmic story

At the beginning of the scientific revolution, a simple cosmic story was based on the early chapters of Genesis, which had been taken at face value. The story was similar for the three monotheistic religions—Judaism, Christianity, and Islam—reflecting their common roots. The great Eastern religions and the animist cultures had different stories. An essential feature of the monotheistic story was that these events had happened in time, albeit a mere few thousand years ago.

As science advanced, this story changed beyond recognition to be replaced by a scientific cosmic story spanning well over ten billion years. Gone were any immediate acts of a Creator producing fully formed aspects of Creation. It was no longer possible to say with Milton,

> *The grassy clods now calved, now half appeared*
> *The tawny lion, pawing to get free*
> *His hinder parts, then springs as broke from bonds,*
> *And rampant shakes his brinded mane* Paradise Lost VII

Instead, what took place was a slow development from the initial formation of the universe so that everything evolved from hydrogen. The skeptic might say, "In the beginning was hydrogen." And so the whole evolutionary story from the Big Bang may be traced, through the heavy elements to the formation of the solar system with the sun and the planets, of life and, ultimately, at almost the last moment, of human beings. At times, this acts as a secular creation story replacing the biblical one. However, regardless of whether it is seen to exclude God, it emphasizes the vastness of time, which is beyond human expectation and comprehension.

The Arrow of Time

Historically, the Christian understanding of time has been linear, starting from Creation and progressing to Christ and the Second Coming—in an "Arrow of Time." In the seventeenth century, the Theorists and Newton followed this linear scheme. Newton's Laws of Motion, which explained and described the motion of the heavens and mundane objects, were an example of a reversible view of time.

Creation was conceived as God making a universe that was up and running rather than one that evolved. Throughout the eighteeenth century this proved to be an excellent scientific way of looking at the universe, enabling the motion of heavenly bodies to be described mathematically

A telescope that belonged to Sir Isaac Newton.

and their paths to be predicted. Similar methods are used today and even Einstein's Relativity and quantum mechanics are essentially reversible. As geologists developed their understanding of deep time and the stratigraphic succession in the nineteenth century, some, notably overtly Christian, geologists—such as Sedgwick, Buckland, and Conybeare—interpreted the succession as a linear process in which God had progressively created various life forms.

In 1830, Lyell gave an anti-progressive gloss to his Uniformitarianism, which prevented him from accepting evolution until 1865. Lyell adopted a steady-state view of the earth, denied directional change in the fossil record, and speculated that creatures similar to extinct forms might recur. Darwin took Lyell's Uniformitarianism and retained the linear geological progression of Sedgwick and others to give a strong directional and irreversible aspect to evolution, often referred to as Dollo's Law.

Mid-nineteenth century physicists began to grasp the concept of the irreversibility of time as the theory of thermodynamics was developed following research on the production of energy from steam engines. The work of Hermann von Helmholtz (1821–94), Rudolf Clausius (1822–88), and Kelvin revealed the significance of entropy (energy still existing but lost for the purpose of doing work, because it exists as the internal motion of molecules) and the Second Law of Thermodynamics. Whatever the physical process is, entropy always increases. Applied to the universe, this means that the temperature will inevitably become uniform throughout. The other implication is that this gives a direction and an irreversibility to the universe, with time pointing like an arrow from the past to the future. From this we get to the "Arrow of Time."

Left behind?

One of the best-selling books in Evangelical bookshops is the work of Creationist author Tim LaHaye. *Left Behind* (1996) portrays the events leading up to the Return of Christ and the Millennium as interpreted by Dispensationalists. Since the mid-nineteenth century, Dispensationalists have developed the Chiliasm of the past into detailed fulfillments of prophecy—some have even suggested that Saddam Hussein is the Antichrist.

Other contenders for this title have been Napoleon and Hitler. LaHaye claims that Christ will return soon and the Tribulation will be inflicted on the godless who are left behind. Not only does LaHaye hold that the earth has little time left, but also that it has existed for only about ten thousand years.

Creationism has become so dominant, especially in the U.S., that many do not realize that it was insignificant until the publication in 1961 of *The Genesis Flood* by Henry Morris and J.C. Whitcomb (scholar of the Old Testament). The main arguments for Creationism are beguilingly simple and reject both evolution and any concept of deep time. While the falsity of Creationism has been exposed on many occasions, its influence remains as strong as ever. It has been presented as "true Christianity" in science education in Kansas in 1999, and in lawsuits in Australia. The Kansas proposals removed all mention of cosmology and deep time, presumably because without deep time evolution is impossible.

Creationism has had a huge influence on popular culture in the U.S. and elsewhere. It is spreading disagreement within the Church and in issues of public education. The National Council for Science Education in San Diego is devoted to opposing this belief. The appeal of Creationism must be recognized. At its heart is a total rejection of deep time and therefore any scientific understanding of the universe.

Creationists, whose leaders often have Ph.D.s in science and engineering, leave most people bewildered about why they adopt such untenable beliefs. Religious brainwashing is not enough of an explanation. Two main reasons are a fear that a rejection of a literal Genesis will erode the Christian faith and a chronological vertigo of the billions of years that geologists and astronomers talk about so glibly and that threaten the finite nature of humanity.

How much time is left?

The end of the world has always held a fascination, if only in disaster movies. Many have had their understanding of the end of the world colored by the predictions of Jehovah's Witnesses or Dispensationalists—views that are not shared by most Christians. There has long been scientific speculation about when the world will end and how long time will last. The two questions are not the same.

For our planet there are two possible fates. In the first scenario, life could be destroyed by a colliding comet or asteroid, similar to the one that struck the earth sixty-five million years ago. But that would not be the end of time as the earth would still exist, even if humans had been destroyed. Then, in about five billion years the sun may become a supernova and explode, destroying the solar system—but there would still be a universe. The final fate of the universe lies much further off and will take place when everything is uniform and entropy has increased to its maximum level. The question is, If there is no more change, can time exist?

Like any development in human thought, the discovery of deep time has not been a simple linear progression as the forces of truth gradually overcame the forces of ignorance. The popular story of the Church holding out for 4004 B.C. as the date of Creation until it was defeated by scientists, is far off the mark.

Before 1600, most people believed that the earth was created in 4000 B.C., simply because this view had remained unchallenged. Significant breakthroughs in the understanding of deep time occurred in the late seventeenth century, with the misunderstood *Theories of the Earth*, and at the end of the eighteenth century, with the simultaneous development of geological understanding all over Europe, the discovery of radioactivity and its application to dating rocks in the 1890s, and the discovery of Hubble's Law in the 1920s, which demonstrated that there had been a beginning to the universe. Only since 1946 has it been possible to date the age of the earth to 4.5 billion years. Since the 1960s, the age of the universe itself has been variously estimated at between ten and twenty billion years. Today, the age of the earth can be dated with fair precision at 4.55 billion years, and the universe less precisely at thirteen billion years, give or take a couple of billion years.

We have come a long way from 1500, when the universe was thought to be six thousand years old. The development of understanding over the last half-millennium has not been smooth, but has progressed in fits and starts. In this process, it is not possible to separate "scientific" understanding from that of the "philosophical" or even the "religious" as all these ways of understanding have had a part to play. At times, these disciplines have retarded our comprehension and at other times they accelerated it. There was no simple liberation of free thought from religious tyranny.

There is now a wide consensus on both the age of the earth and of the universe, with a knowledge that not only the universe has a beginning, but so does time. We are now back with Augustine who claimed that the universe was not created *in time* but *with time*. This is the question that cosmologists such as Stephen Hawking (b. 1942) continue to grapple with. Yet, despite the irrefutable scientific evidence of vast antiquity, a sizeable minority of the Western world still rejects deep time in favor of a universe that has been in existence for only a few thousand years.

PLATFOR

10: more clocks within us

John Wearden

Psychological perceptions of time

Time spent waiting for a train passes excruciatingly slowly. The pace seems to revive when the train pulls in to the station—so what are these clocks inside us?

Never does a minute seem so long as when we look at the seconds hand moving round the face of a watch or clock G. J. Whitrow

There seems to be something intangible about time: unlike light or sound, it does not seem to be a thing in itself but rather a property of events that take place within it. Some have even said that "events are perceived, but time is not." The problem of how time is perceived is further complicated by the fact that there is no obvious organ for time, no equivalent of the eye or the ear, the physiology of which offers a starting point for understanding how light or sound is perceived. Nevertheless, psychologists and physiologists have proposed that specific mechanisms for perceiving time do exist.

Telling how long it takes

The principal area of interest for time psychologists is the perception of duration—of how long events seem to last. Noticing duration is a crucial part of our ability to perform commonplace actions such as holding a note the right amount of time when singing, or doing something after the right amount of time has passed. Attempts to understand time perception by careful experimentation in the laboratory began in the late nineteenth century, mainly in Germany and the U.S. Experiments involved people judging how long events appear to last, or which of two events lasts longer. Experimenters aimed to find out, for example, what people did when they were asked to hold down buttons for specified lengths of time.

Many of the early results remain valid today. One question was what happens when the duration of a sound and the duration of a visual stimulus (illumination of a light bulb, for example) are compared. When in fact they have the same duration, the sound is judged longer than the light, usually about 10–20 percent longer. A similar effect, but one that is if anything even more marked, is the "filled duration illusion." Suppose we compare the perceived duration of two equally long events. One is "filled"—for example, a tone that comes on and stays on for some time. The other is "empty"—such as an interval starting and ending with a very brief click. The filled interval will be perceived as longer than the empty one, often as much as 30 percent longer.

Twentieth-century psychologists turned to the idea of an internal clock to help explain these observations—some internal process by which people are able to record and add up passing moments. If such a mechanism actually exists, it will be biological, and neurophysiology will have to advance greatly before it can study it directly. However, psychologists have found it possible to make predictions about some aspects of how such a clock would work, and to test their predictions by experiment.

Time hot and cold

The idea of an internal clock is always associated with the name of the physiological psychologist Hudson Hoagland, whose work was published in the 1930s. Hoagland's theory was stimulated by a real-life incident. His wife was suffering from influenza and her body temperature was raised by a fever. Hoagland was caring for her but took a short break away from her bedside. On returning, he was berated by his wife for an exceptionally long absence, when in fact he had been away for only a short time. For some reason, the time elapsed had seemed longer to her than to him, and longer than it was in reality. But why? Hoagland's interest in physiology led him to conceive the idea of a "chemical clock"—that is, some chemical process in the brain or body that is used to judge the passage of time. By the laws of physical chemistry, the speed of any chemical process will increase when it is heated up, so Hoagland reasoned that his wife's clock—heated by her raised body temperature—would be running exceptionally rapidly, thus making her think that a long time had elapsed when in fact it had not.

Hoagland then did to his wife what any self-respecting 1930s scientist would do: he experimented on her. More specifically, he asked her to count up to sixty at a rate of what she judged to be one count per second. He measured the time she took and her body temperature as this rose and fell during the days she was ill. In general, the hotter Mrs Hoagland was, the faster she counted: exactly the result predicted by the "chemical clock" theory.

Unknown to Hoagland, a French psychologist called Michel François had published a similar result a few years earlier. François heated up his experimental subjects using diathermy, the passage of high-frequency electric current through the body (a procedure disturbingly reminiscent of the operation of a microwave oven), and found that a higher body temperature seemed to make some internal timing process run faster. These two sets of results gave rise to what must be one of experimental psychology's most bizarre fields of research, that of the study of the effects of body temperature on subjective time. Manipulating body temperature is difficult (and can be dangerous, so ethical considerations prevent modern replications), and many different methods (heated rooms, heated helmets, natural fevers as in Hoagland's study, and tanks of cold water) have been used to accomplish it. In spite of many differences between studies, a recent review found that the bulk of the data went in the direction of Hoagland's initial idea: some internal timing process seemed to run faster when people were hotter.

Accumulating time

The next significant breakthrough in time psychology came with the development of precise mathematical models of the internal clock in the 1960s and 1970s, with the names of the psychologists Michel Treisman and John Gibbon being particularly prominent. These models do not depend on particular biological processes by which the clock might work (for example, whether they depend on chemistry). The models strip the idea of an internal clock down to the bare minimum, leaving it with some form of pacemaker only, which can be anything at all that can produce something countable (such as pulses or ticks), some form of accumulator (which can be anything at all that records how many pulses or ticks have been produced), and a switch-like connection (that can be switched on or off) between the pacemaker and the accumulator. The onset of an event to be timed causes the switch to close, allowing pulses from the pacemaker to be transmitted to the accumulator, much like the closing of an electrical switch allows current to flow. When the stimulus goes off, the switch opens again, so the content of the accumulator is the number of "ticks" of the internal clock that have accrued while the event was going on—in effect a raw measure of duration.

This simple model of an internal clock is surprisingly powerful. One result is perhaps initially somewhat surprising. It seems that, on average, real time and subjective time are the same. That is, when we record the

The idea of a simple pacemaker-accumulator clock in the brain is shown here. The pacemaker is a mechanism that produces pulses or "ticks." These are accumulated in the accumulator when the switch is closed. The switch acts like a light switch. It closes to allow the pulses to flow, then opens to stop the flow. The switch might close when some stimulus to be timed starts, and open when it stops, so the accumulator contains the number of ticks that have occurred during the stimulus, a kind of "raw" measure of its duration.

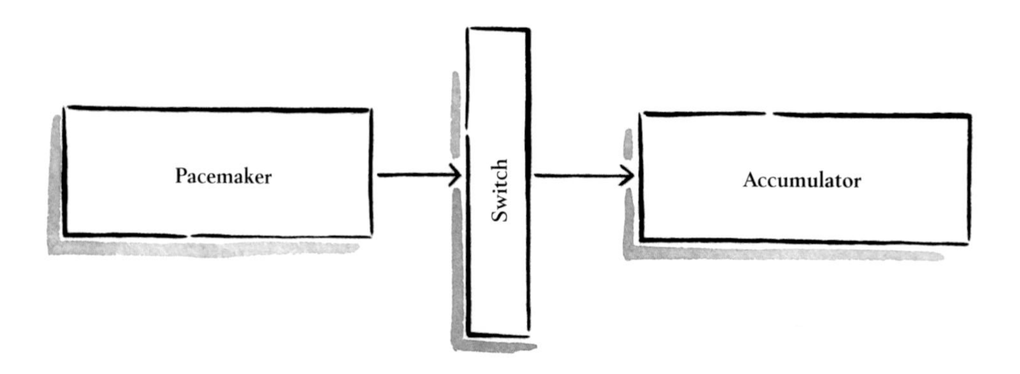

judgments that people make about various real-time durations—those measured by mechanical or electronic clocks—the judgments are found on average to be almost accurate.

It might seem miraculous that such a simple relation between real and estimated time occurs, but in fact it arises naturally from many sorts of accumulation process, of which the ticking pacemaker-accumulator clock is only one example. Another is a simple pacemaker-accumulator system consisting of a running tap that fills a measuring jug. We use the amount of liquid in the jug to measure the duration of various events, by turning the tap on when the event starts and turning it off when the event stops. Suppose that the time taken to start and stop the tap is negligible, and that the flow from the tap is constant. If the duration of the event is doubled or halved, then the amount of liquid in the jug is also doubled and halved.

This assumes, though, that the switching process (turning the tap on and off) takes a negligible amount of time. Consider next the case in which turning the tap on or off takes time. Obviously, the accumulation of water is not exactly proportional to real time, but it still may be fairly close if the time taken to open and close the tap is small compared to the duration of the event being timed. In other words, turning the tap on and off will make a bigger contribution to the measurement of short times than of long ones. This is exactly the prediction of internal clock theory, and the prediction is supported by experimental results.

The tap-and-jug metaphor also illustrates another puzzle. Suppose the flow from the tap (or the rate of ticks from the pacemaker) is not constant and that the measurement is repeated in a number of trials, then the measures are averaged together. A pacemaker that did not tick at a constant rate, or a tap that did not flow at a constant rate, could be variable in two ways: it could either change rate within trials, or between trials (being constant within trials). In the first case, the tap flow would fluctuate randomly from moment to moment around some average flow rate. In the second case, the flow rate would be constant on one trial, but have a randomly different constant rate on the next. Although what happens is not as immediately obvious intuitively as when the tap flow is completely constant, mathematical analysis shows that the result on average is the same: doubling or halving the real time will, on average, double or halve the amount accumulated in the jug.

The "on average" is important here, because the result now holds only when the results from a number of trials are averaged together. If the tap flow varies within or between trials, the amount accumulated in the jug will not be exactly the same from one trial to the next. In other words, variation in flow affects variability in time judgments from one trial to another, but does not affect the average, which remains proportionally related to real time.

The tap-and-jug metaphor illustrates why the simple relation between subjective and real time is so natural, and how it can arise from accumulation processes where we don't necessarily need to assume that the rate of ticks is constant. To distort the proportional relation we would have to make some additional assumption, such that the rate of ticks was slower or faster depending on the duration timed, but there seems no obvious reason for doing this. In general the simple proportional relation between subjective and real time found in experiments is not only the most natural relation possible, but is also one compatible with a simple pacemaker-accumulator model.

Speeding up and slowing down

Another situation of interest is one where the rate of the internal clock is systematically changed. Hoagland's work offers one technically difficult way of doing this, but certain drugs and other manipulations may also change pacemaker speed. In general, suppose something is done to change the speed of the internal clock. What are the effects on subjective time judgments? Intuitively, a faster clock sends more ticks or more water to the accumulator or water jug, but in fact the effect is more specific than that. The resulting accumulation (of clock ticks, or water in the jug) is actually the multiple of clock speed and real-time duration, so effects of heightened clock speed rate are greater at longer intervals than shorter. Such relations are used as evidence of a change in clock speed in modern research.

However, the effects of changing internal clock speed are more complicated than they might at first seem. A central problem is the problem of comparison: if the speed of the internal clock is changed, what is this change compared with, and how is the judgment assessed? Let's take two different examples.

In the first, a person has to judge the duration of an event. During initial training, the person will probably unconsciously associate some number of internal ticks (say fifty, but the number is entirely imaginary) with a duration of one second, so the normal clock speed is fifty ticks per second. Now the internal clock is speeded up in some way, so it makes sixty ticks per second. An event that really lasts one second will now be estimated as 1.2 seconds long (as 60/50 = 1.2). In general, increasing pacemaker speed makes the duration of external events seem longer than "normal."

In the second example, the person is trained to produce a one-second interval (for example, by holding down a button). In the normal state, this duration is associated (let's say) with fifty internal clock ticks, but then the person is asked to do this again with a speeded up internal clock that produces sixty ticks per second. The person counts fifty ticks

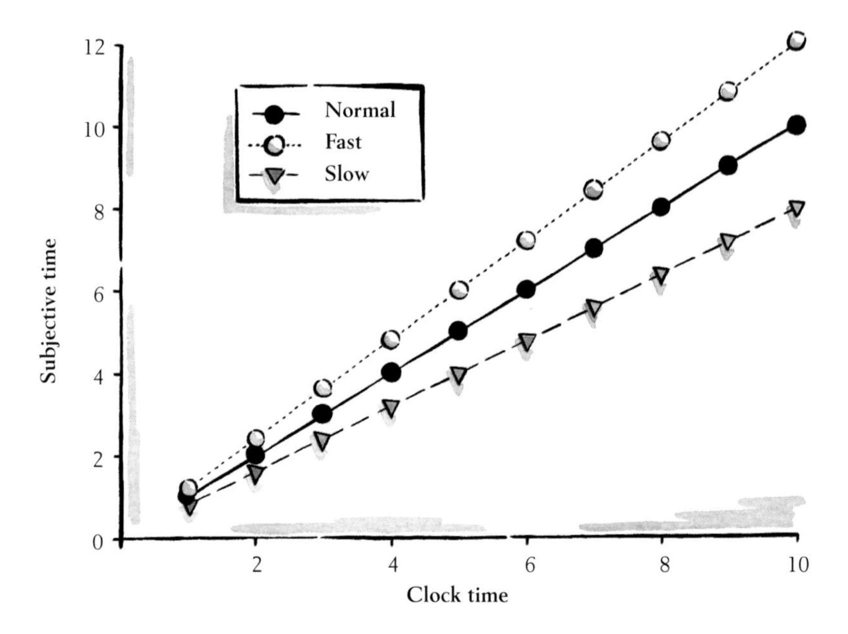

The effect of a change in pacemaker speed is greater at longer times (for example, ten minutes) than shorter ones (such as two minutes). Stimuli lasting from one to ten minutes are presented on the graph. The pacemaker runs at Normal speed, or 20 percent faster than normal (Fast) or 20 percent slower (Slow). The graph shows the effect on subjective time.

then responds, but with the faster clock the fifty ticks take only 0.83 seconds to accumulate (50/60 = 0.83). Increasing clock speed shortens the duration produced.

So, the prediction is that some manipulation that is supposed to increase or decrease clock speed will have opposite effects on the estimation of the duration of events and production of specified intervals of time, and this is the result obtained.

Getting wound up

One factor often suggested as a determinant of internal clock speed is the arousal level of the person. A fairly common anecdote describes changes in subjective time in life-threatening situations. For most people these days, car accidents are by far the most common type, but soldiers in battle often give similar accounts. The stories are generally consistent: in a high-threat situation—in the seconds before the collision in a car accident, for example—external events seem to slow down (they seem to last much longer than normal), and people report seeming to have lots of time to perform actions (such as looking in the

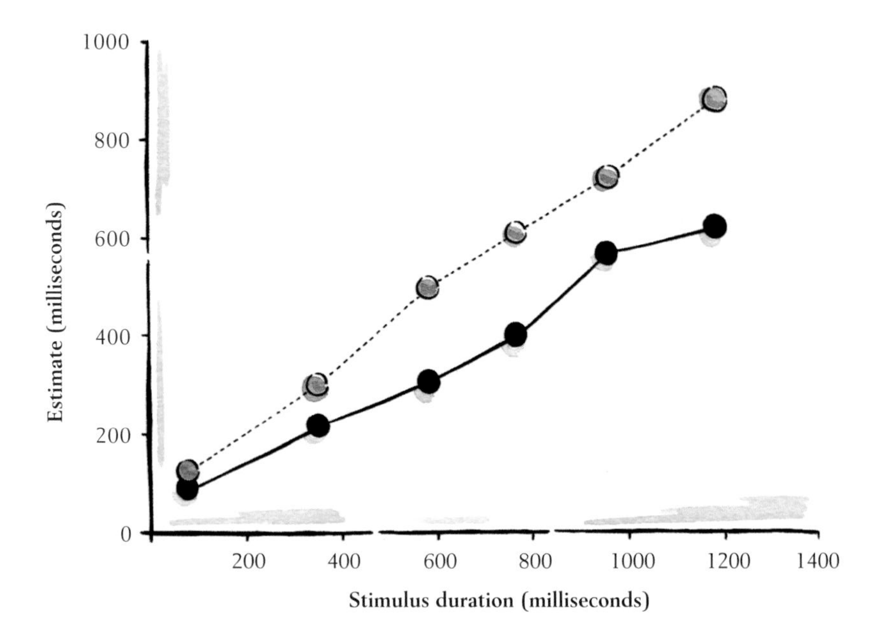

Verbal estimates of the duration of a number of tones (auditory stimuli) or squares presented on a computer screen (visual stimuli). The actual durations ranged from seventy-seven to 1,183 milliseconds (thousandths of a second). Estimates of the auditory stimuli are shown as light-colored circles; estimates of visual ones as dark circles.

mirror, checking the seat-belt, and putting the handbrake on) in what objectively must be a very short space of time. Such effects are exactly those obtained if the speed of the internal clock is vastly increased. Now, external events will appear to last longer than they normally do—that is, external time seems to "slow down." Conversely, reactions may speed up, and this type of change in timing has obvious survival advantages; people have "more time" to react to the slowed external events, and their reactions are perhaps speeded also. However, for obvious reasons, life-threatening situations cannot be arranged in the laboratory, and attempts to demonstrate that relatively small changes in arousal change the speed of an internal clock produce mixed and complicated results.

On a less dramatic note, the simple pacemaker-accumulator clock model may also offer us insight into the time distortion effects mentioned earlier, obtained when sounds and lights, or filled and unfilled intervals, are compared. Suppose that the difference in these cases is simply a small difference in clock speed: the pacemaker, for some reason, runs slightly faster for sounds than for lights, or for filled than unfilled intervals. The prediction here is that the difference between the compared conditions should increase as durations become longer (remember that accumulator contents are clock speed multiplied by duration), and this is exactly the result obtained.

Overall, therefore, the pacemaker-accumulator model that was developed in the 1960s, based on a simple and natural accumulation process, not only makes quite precise predictions about timing behavior, but also can deal in a straightforward way with some time distortion effects that have been puzzling for a hundred years or more. However, alternatives to internal clock models exist, particularly when timing in "real-life" situations is involved.

Times remembered

Internal clock models have had great success in explaining certain laboratory-based situations, particularly those where the person is alerted by instructions that time judgments are going to be made (for example, "I want you to hold down this button for one second" and "You are going to hear two tones, and I want you to say whether the first or second one lasted longer"). However, unexpected questions about elapsed time can also be asked. Here's one. How long has it been since you began to read the page that this sentence is on? Whether your answer is accurate or not, it is certainly not random, in the sense that you would never say "a few tenths of a second" or "three hours" (unless you are an exceptionally fast or slow reader!).

How is the time judgment performed? If you didn't know that the question about time was going to be asked, how can you have started your internal clock, and what constituted the "starting point" for the clock? It seems that some other mechanism must be used, and theorists have generally proposed that some sort of memory or event storage is the critical factor.

In the 1960s, Ornstein proposed that human timing was not based on an internal clock at all, but rather on the accumulation of memories during the time period to be judged. So, for example, when the unexpected timing question is asked, you retrospectively search your memory, and the "amount" of memory found there determines the time judgment. The rule that Ornstein proposed for this is "storage size"—that is, if you find a lot stored in your memory, the time elapsed is judged as longer than if little is stored. Today, most psychologists would accept that some timing in humans is accomplished by an internal clock-like mechanism but that many time judgments, perhaps even the majority of those carried out in real-life situations, are determined by some sort of memory-based counting of events.

Ornstein's position is supported by experiments which show that time intervals with more "psychological activity" (for example, more things to be

Journal des Voyage

ET DES AVENTURES DE TERRE ET DE MER

(SUR TERRE ET SUR MER; MONDE PITTORESQUE; TERRE ILLUSTRÉE réunis)

DIMANCHE 5 JANVIER 1902

Journal hebdomadaire. ABONNEMENTS : UN AN : PARIS, SEINE ET SEINE-&-OISE, 8 fr. — DÉPARTEMENTS, 10 fr. — UNION POSTALE, 12 fr. *rue Saint-Joseph,* 12, P

N° 266 2e SÉRIE

LES ALPES HOMICIDES

TERREURS D'HIVER

PAR JULES LERMINA

PR 15

Souffleté par cette avalanche, le malheureux s'élançait en avant, criant des clameurs d'épouvante. (P. 92, col.

QUATRIÈME NUMERO DU JUBILÉ

remembered, more events to be perceived and judged) tend to be thought of as longer than those with less activity. Some research suggests that drugs which increase arousal make intervals seem longer when we are told in advance to time them, but have no effect on remembering a past quantity of time. Drugs that affect memory storage influence the latter but not the former. However, some results suggest that a simple "storage size" account of timing is much too simple.

The waiting room

Anecdotes from two familiar situations illustrate two slightly different problems. The first is people's experience of the passage of time while in some very boring situation, such as waiting for a train or plane. In this "waiting room" effect, virtually everyone reports that time seems to drag, and that each minute seems to last an exceptionally long time.

Why? Since nothing much is happening, there's hardly anything to store in memory, so time should be judged as very short rather than long, which it obviously is not. The second situation is in some ways the converse of the first. You are engrossed in an activity (such as intense study, watching an engaging film or play, or playing a demanding computer game) then you observe that a considerable time has passed without you "noticing." This effect is even enshrined in the phrase "time flies when you're having fun." In this second case, the intensity of the activity you've been engaged in should have resulted in a considerable amount of memory storage, causing the interval to be judged as abnormally long, rather than short.

These situations are actually very complicated in terms of the psychological processes going on in them, and it seems likely that at least part of the effect obtained depends on when the time judgment is made. For example, suppose we contrast a period spent in a waiting room with that spent watching an exciting film. Some time after the period has ended, we ask how long the different situations lasted. Now, the results may not contradict the storage size model at all; the duration of the film may well be judged as much longer than the period in the waiting room. The contradiction to the storage size model actually comes when the judgment is made while the person is in the situation. How can this be explained?

The answer is generally thought to involve attention to time so, for example, time in the waiting room seems to drag as, in the absence of other things to do and when we are waiting for something which is going to happen at a particular time, attention is focused on time, which then seems to pass slowly. Conversely, during the exciting film, attention is not focused on the passage of time but rather on the film's contents, so the time passes quickly.

ime appears to be tanding still for this an, lost in a violent nowstorm. His sense of irection gone, death tares him in the face nd he relives his whole fe in an instant.

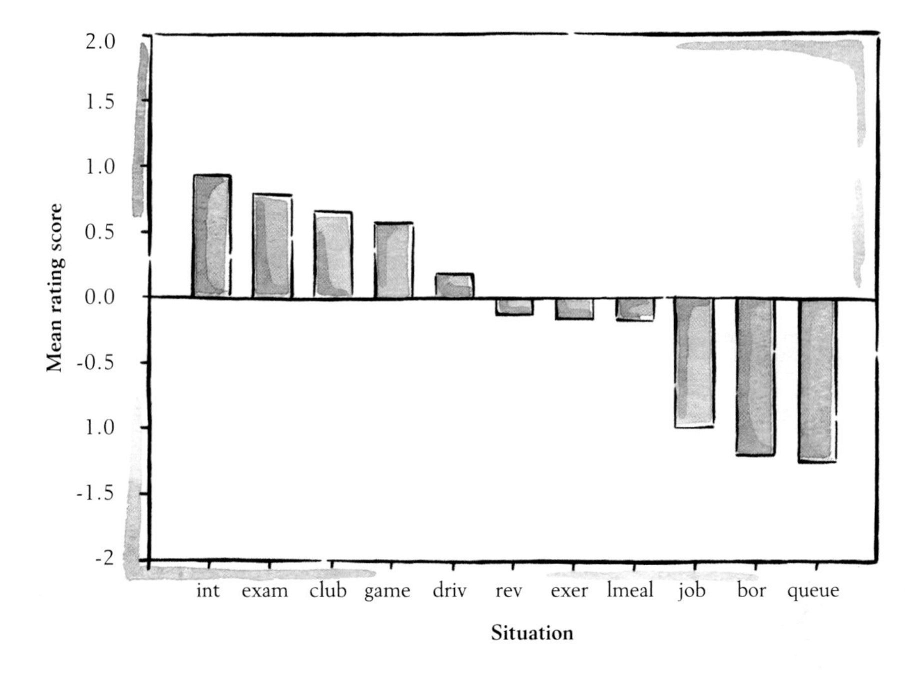

Average ratings of the passage of time in different situations. Sixty students were asked whether time appeared to pass normally (score of zero), faster than normal (positive scores), or slower than normal (negative scores), in different situations. From left to right in the figure, the situations were: watching an interesting film (int); in an examination (exam); dancing at a club (club); playing a computer game (game); driving (driv); doing examination revision (rev); doing exercise (exer); eating a leisurely meal (lmeal); working at a job (job); watching a boring film (bor); waiting in a queue (queue).

In a recent study in Manchester, sixty students were asked to say how long, compared to some imagined "normal" state, various events lasted. It is obvious that "exciting" events where little attention is being paid to time are judged as passing more quickly than more boring ones, consistent with the idea that the amount of attention paid to time is an important influence on the subjective passage of time, when events are actually going on.

Even in experimental situations where the preferred explanatory mechanism for time perception is an internal clock, attention may still play a role. For example, suppose you are required to judge the duration of an event in two conditions. In one of these, timing the event is all you have to do. In the other, an extra task must be performed along with the judgment of time. In this case, a number of studies suggest that the extra task will make the timed event seem shorter. It is as if the extra task distracts attention from the timing mechanism, so that ticks from the internal clock are missed.

…e same period of time …n pass slowly for a …red person, and flash …st for those who are …npletely immersed in …eir activity.

A popular idea is that timing and non-timing tasks compete for a limited pool of attention; the more attention paid to the non-timing task, the more likely it is that the accuracy of the timing will be affected. An easy non-timing task needs little attention, so hardly affects the time judgement at all, while a difficult non-timing task commands attention and timing is adversely affected, usually in the direction of events being perceived as shorter.

Overall, the relations between memory, attention, and time judgments are very complicated, with theories dealing with their interaction often being vague or even implicitly contradictory. We have already seen how the idea that "more stored events = more time" can work in some situations but is contradicted in others. A further problem is that the events that are supposed to be stored to form the basis of time judgments are difficult to specify. Internal clock models at least have the advantage that the operation of the pacemaker, switch, and accumulator can be specified in mathematical form, so that exact predictions about behavior can be made. These predictions may sometimes be wrong, but at least they are (usually) clear. Memory-and-attention models of timing, on the other hand, are necessarily more complicated, and only rarely lead to exact mathematical predictions about behavior. However, there is little doubt that many real-life timing phenomena are more consistent with the idea that memory storage or attention, rather than an internal clock, is used to generate time judgments. It seems likely that a more complete judgement will require both an internal clock and a memory-based "event accumulator," and the construction of such hybrid theories is currently under way.

Timing in children

Until recently, the study of timing in children proceeded quite independently of research on adults. It has been dominated by the well-known Swiss developmental psychologist Jean Piaget (1896–1980), who was concerned with how children put together their knowledge of time. Compared with the standard of the experiments on adults, the experimental situations used on children are very complicated.

Something of particular interest to Piaget was the development of correct judgments of relations between distance, speed, and time when moving objects (toy trains or cars were most commonly used) were observed by children. A typical situation is shown in the diagram. Here, the travel times of two toy trains on different tracks must be compared. The trains may start together or not, end together or not, traverse different or the same distances, and travel at the same or different speeds. It is easy to imagine that ascertaining which train would run for longer can be difficult even for adults, when all these factors are varied. Fortunately, Piagetian researchers have concentrated on some relatively simple cases—for example, ones in which the trains start together but end

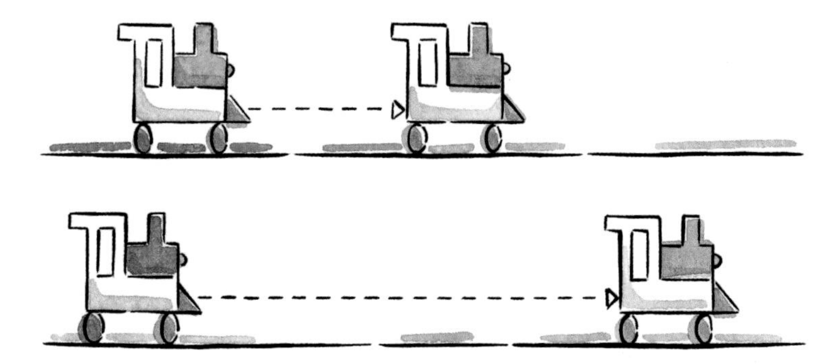

Piagetian time/space problem. The two toy trains are on different tracks, but both can be seen by the child at once. Each train moves the distance shown in the dotted line. They may start at the same time, or different times, go different distances, and travel at different speeds. The child's task is to say which train ran for the longer time.

at different times, after traversing different distances (which, of course, entails traveling at different speeds).

Even young children master judgments based simply on the different end times of trips made by the trains, and conclude that, if the trains started together, the train that finished last ran for the longer time. When start and end times are not the same, the problem becomes more complicated, and children may "centrate" (to use Piaget's term) on the start or end time, but not both, and may fail to coordinate the two correctly. For example, they may conclude that the train that started first ran for longer (regardless of its stop time) or that the train that stopped second ran for longer (regardless of its start time). The term centration here refers to the child's fixation on one aspect of a situation when more than one (start and end time in this case) is needed to answer the question posed.

The distance and speed of the train's travel are also relevant to time judgments and here young children are posed a particularly thorny problem, as the relations between travel time and distance, and travel time and speed are not the same. If the speed is constant, greater distance means greater time—that is, the relation between distance and time is a directly proportional one. In contrast, if the distance traversed is constant, the relation between time and speed is an inversely proportional one (faster speed means less time). The child may centrate on either distance or speed, and use the same, simple, directly-proportional relation for both. This means that the child concludes in one case, usually correctly, that greater distance means more time, or concludes, usually incorrectly, that faster speed also means more time. With increasing age comes increasing cognitive flexibility, so children will eventually learn to coordinate start and end times correctly and, finally, even to appreciate the correct physical relations between time, distance, and speed, involving

both the direct and inverse proportionality, so that time = distance/speed.

Another issue studied by time psychologists in the Piagetian tradition has been a child's developing appreciation of change over time. For example, adults know that as a tree grows it changes not only quantitatively (gets bigger over time), but also changes qualitatively (it grows more branches and leaves). Young children, on the other hand, may illustrate development over time in their drawings only quantitatively. That is, if given a drawing of a newly planted sapling, and asked to draw what it looks like years later, young children just draw the same thing, but bigger. Older children coordinate an increase in size with changes in shape.

While the majority of existing work on the developmental psychology of time comes from the Piagetian tradition, a recent trend is to study with children the simpler experimental situations used with adults. So, for example, children's discrimination of the duration of simple events like lights or tones has recently been a focus of interest, as have the effects of distraction.

The latter area produces some unusual findings. For adults, as has been mentioned above, having to perform on a non-timing task at the same time as making a time judgment almost always shortens the duration judged, as timing and non-timing processes compete for attentional resources. With young children, on the other hand, the distinction between timing and non-timing processes may be more poorly developed, so all sorts of information, whether relevant to timing or not, are lumped together and children judge that "more events means more time." For example, adding a non-temporal task to a time judgment shortens the judged time in older children, as in adults, but lengthens it in younger ones, as if the children do not distinguish appropriately between temporal and non-temporal information.

Time and the brain

All psychological processes involve brain activity, but this does not mean that models conceived in purely psychological terms, like the pacemaker-accumulator clock or the event storage models, cannot significantly advance our understanding. Simple psychological models may be more useful than complicated brain models, particularly when many aspects of brain function remain obscure. However, there has been much recent research focusing on the issue of how the brain might perform timing tasks, either by searching for the neural basis of some sort of internal clock, or by other means.

The human brain is a highly complicated object, possibly the most complex that exists. It consists of billions of neurons, arranged into hundreds of distinctly identifiable nuclei (and possibly hundreds more that have not been clearly distinguished as yet), with myriad interconnections

at is the time,
. Policeman? by
rman Hepple.

between them. Given this complexity, it is perhaps not surprising that research on how the brain performs timing is still in its infancy. Nevertheless, some definite facts have emerged, and modeling how the brain might mediate timing tasks is currently a lively topic.

A convenient starting point is to examine some apparently simple drug effects. The neurons of the brain communicate with each other chemically using various neurotransmitters, and different regions of the brain consist of neurons that use different neurotransmitters. Furthermore, the different neurotransmitters may, at least to some degree, be specific to certain types of psychological process. One important neurotransmitter in the brain is dopamine, and although dopamine neurons in different parts of the brain have diverse functions, a common effect of changes in dopamine levels, at least in humans, is changes in arousal or mood. Indeed, many common anti-depressant drugs act by increasing dopamine levels, and a lack of dopamine in certain regions of the brain is thought to be linked to some sorts of depressive illness.

There is considerable evidence that dopamine levels in some regions of the brain play a critical role in timing, although there is controversy as to exactly what that role is. The relevant findings are most clearly illustrated by reference to experiments on animal timing. At the start of a typical experiment, a mildly hungry rat is trained to press a small lever and is rewarded with small pellets of food for doing so. The experimental session is divided into trials, and between these trials the lever is withdrawn and the animal cannot respond. A trial starts with the illumination of a light, then the first lever-press occurring a certain number of seconds (say, twenty seconds) after the start of the trial is rewarded with food. Food delivery stops the trial, and another one is given later, and so on. The rat can press the lever any time it wants to during the trial, but eventually, the animal will start pressing some time just before the twenty seconds (for example, at fifteen seconds). When the animal is well trained, so-called "peak trials" are introduced. Here, no food is given, and the trial (which starts with a light going on as usual) lasts much longer than the time usually associated with food (twenty seconds in our example). In peak trials when the rat's response is averaged over a number of such trials, it peaks almost exactly at the time when food is usually available.

Suppose that the rat has been trained with a twenty-second food time in a "normal" (undrugged) state. Now the rat is given some trials after it has been injected with amphetamine, a drug which increases dopamine levels. The rat's peak response will shift to the left (for example, from twenty to eighteen seconds). The task described is rather analogous to the rat producing a time interval, and this shortening of the time produced is an indication of the internal clock being sped up. With a faster clock, the person or animal responds earlier than before.

A similar experiment involves training the animal, then injecting it with haloperidol, a drug which decreases dopamine levels. Now the rat's peak response is shifted to the right (to, say, twenty-two seconds from twenty). This result is consistent with the haloperiodol having slowed down the internal clock.

It seems therefore that a drug that increases dopamine levels makes the clock run faster, and one that decreases them makes it slower. This result is also supported by the rat equivalent of experiments on estimation of the duration of events. Some dopamine system in the brain seems implicated in timing in rats, and experimentally induced brain lesions that eliminate dopamine in some brain areas may result in animals that show no sensitivity to time at all: they are unable ever to learn that food comes at a certain time, and unable to respond to stimuli on the basis of their duration.

However, other neurotransmitter systems do other things. Drugs that affect acetylcholine-based neurons, which are believed to be implicated in memory storage, do not change the speed of the internal clock, but instead seem to change the time that the animal learns. So, for example, if the time associated with food is twenty seconds, the animal remembers this as twenty-two or eighteen seconds, and does not peak its responding at twenty seconds. Such memory effects can be distinguished from clock-speed effects by experiment.

Humans and drugs

Drugs have also been used to study timing in humans. For various reasons, direct analogs of the rat experiments cannot be performed, so results are less clear. There is some evidence from humans that dopamine levels affect internal-clock speed, but results are much less clear than in rats. Likewise, changing acetylcholine levels in humans may affect memory for time, but it also affects memory for other events, so alters time judgments based on the number of events remembered.

Some patient groups are interesting from the perspective of the neurotransmitters involved in timing. Patients with Parkinson's disease have depleted dopamine levels in the very brain areas (basal ganglia) that are implicated in timing in rats. Ordinarily, patients take drugs (usually l-dopa) to stimulate their dopamine levels artificially. Do Parkinson's disease patients have timing deficits? Initially the answer seemed to be unequivocally yes. Patients received training to tap along with a metronome, then were required to continue the tapping at the same rate when the metronome stopped. Patients showed clear deficits compared with non-patients in terms of the accuracy and variability of their tapping, and were also worse when they were off their l-dopa medication that normally increases dopamine levels. However, Parkinson's disease is primarily a disease of motor behavior—that is, a disorder of movement—so it is perhaps unsurprising that a task like repetitive tapping is affected.

What happens when patients are tested on the discrimination of the duration of stimuli, where movement requirements are minimal? Recent laboratory research has shown that patients have no deficit at all—or at best a very subtle one—suggesting that stimulus timing processes, rather than timing processes which control movement, may be much less dependent on dopamine in humans than has previously been thought.

Brain scanning is another method recently used to explore timing processes in the brain. Not only can the traditional electroencephalogram (E.E.G.) record the electrical activity of the brain from the scalp, but techniques like positron emission tomography (P.E.T.) or functional magnetic resonance imaging (fM.R.I.) can give measures of the relative activity of different brain regions when different sorts of tasks are performed. Do such scanning studies tell us where the internal clock is?

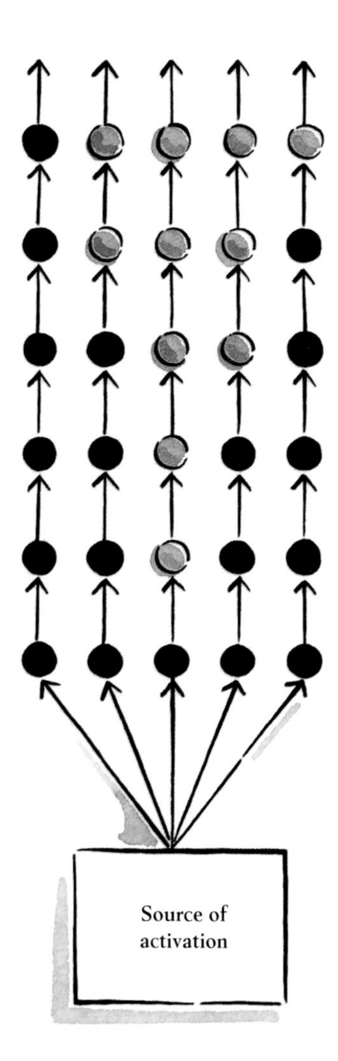

Timing by spreading activation. A "source of activation" like a pacemaker repeatedly sends "pulses" through a neural net. When a unit is activated (filled), it can pass on its activation to the next unit in the chain when it next receives a pulse, but the probability of passing the activation on is less than one, so the activation doesn't necessarily always increase. On average, however, as more time passes, more of the neural net is activated, so the number of units activated can be used to measure time.

The results of experiments are not so clear and many areas of the brain seem to act together when a timing task is performed. This is perhaps not very surprising: when a stimulus to be timed is presented—a tone, for example—it also has other sensory characteristics as well as its duration (such as pitch, loudness, timbre, and location) and part of the brain activity will be concerned with these rather than time as such. Another problem is that to make a time judgment (such as whether a tone has the same duration as one that was heard a while ago) involves memory and decision processes—for example, the retrieval of the remembered duration and comparison with the current duration. All these processes will also show up on scanning, so isolating the brain processes solely responsible for timing, if such exist, remains a difficult problem but one which currently stimulates a lot of research.

A rather different problem, also unsolved, is how brain mechanisms might actually do timing at all. One problem is that neurons operate very rapidly, often firing repeatedly every few hundredths or thousandths of a second. If these neurons represent the basic ticks of some internal clock, then how are such rapid ticks accumulated to provide representations of times which are sometimes tens of thousands of times longer? For example, in a peak procedure experiment, a rat may learn that food comes 120 seconds after the start of a trial and peak its response then. How can such a long time be represented by neurons firing fifty times per second?

One way is to assume that the neuronal activity is counted somehow by the brain (which must, of course, be able to count to quite a high number to represent 120 seconds with ticks that occur fifty times per second). Various sorts of neural network models have been proposed to do this. One possible idea is that the ticks from some neural internal clock spread activate a network of neurons in a manner rather similar to the way that dropping a stone into a pond "activates" the pond.

In the diagram on the left, the number of neurons activated in the network is a measure of time, just like the area of the pond "activated" by a stone is a measure of how long has elapsed since the stone was dropped into the pond. Such a mechanism seems potentially plausible, but no model of timing that uses this idea can, at the time of writing, simulate behavior on real timing tasks very well.

The study of time and the brain is still in its infancy, and is being actively pursued using a mixture of techniques: drug studies with animals and humans, studies of patients, scanning studies, and computer modeling. Perhaps future investigations will suggest that our metaphor of an internal clock, useful though it has been, is actually incorrect in that it does not correspond to any brain mechanism, and that the brain performs timing tasks by some process as yet unimaginable.

11: puzzles about time

Robin Le Poidevin

Points to ponder

...But to apprehend
The point of intersection
of the timeless
With time, is an occupation
for the saint T.S. Eliot

The greatest puzzle about time is what time is. How is it defined? What is its nature? We say that clocks measure time, but cannot point to what they measure, as we can point to the kinds of things that rulers measure. What is certainly true is that we become aware of time by noticing changes, either in the environment around us or inside us: in our bodies or our thoughts. So perhaps that is what time is: change. But time is not just the changes we perceive; there is another change, perhaps more fundamental, and that is the passage of time. Future things, such as a long-awaited reunion, become present and then ever more past. We could picture this as a stream of events moving through time. And this passage always seems to be in one direction. We can never revisit a past scene. Time also never stops, or even slows down or speeds up, although it may sometimes appear to do so.

What is time?

Simple though they are, these reflections on the nature of time provide a source of deep problems that have exercised thinkers since ancient times. In this chapter, we shall take a brief look at a number of such problems, some of which suggest that time itself cannot exist, or at least that our ordinary conception of time must somehow be at fault. It is commonly known that, in the twentieth century, physics brought about a revolution in thinking about time and space, but the puzzles presented here are ones which stem not from advances in science, but from reflection on the ideas themselves. Although some possible responses are indicated in some cases, no attempt is made to provide a list of definitive solutions because unlike, arithmetical puzzles, they do not have answers that everyone would accept as the correct answer. Do they have answers at all? That is for you to decide. Some may strike you as trivial, or based on obviously faulty reasoning. But in thinking about them, you are led to reflect on the nature of time itself. The following puzzles should be seen, then, as starting points for your own thoughts about what time really is.

The arrow

Imagine an arrow in flight. It is moving for the whole of the time between leaving the archer's bow and hitting its target. Since it never pauses mid-flight, that ought to mean that it is moving at every moment during its flight. But now consider what a moment is. In particular, consider how long a moment is. A second? Half a second? Whatever figure you come up with, you can always imagine a smaller one. So every "moment," if it takes up any length of time at all, will have smaller moments within it. But there must be a smallest moment. How big would that be? Presumably, it would take up no time at all; an instantaneous moment, rather like a point on a line. Just as a point takes up no space, an instant takes up no time. So is it true that the arrow moves at every instant during its flight? No, for nothing can move in no time at all, and an instant we have just defined as taking up no time at all. So if nothing can move in the space of an instant, how can it be said to be moving at that instant? But then if we say that, at every instant of its flight, the arrow is not moving, it seems to follow that the arrow is not moving for the whole period of its flight. In other words, the arrow never moves at all!

The choice of an arrow is a completely random one; we could have chosen any moving object. But then if the reasoning above is sound, it follows that all motion is impossible. And we can extend this conclusion still further, for motion is just one kind of change. Other changes include our changing shape as we grow older, the cooling-down of a cup of tea, the ripening of a strawberry, the burning of a candle. The paradox of the arrow applies to all of these. In an instant, there can be no change whatever. So nothing can change in a period, for such a period just consists of a number of instants. But if the paradox of the arrow shows that motion is impossible, it also shows that all change is impossible. And if time is the same thing as change, then time must be unreal.

Zeno of Elea

There are two famous Zenos in the history of philosophy. One was Zeno of Citium (336–264 B.C.), who founded the philosophical movement known as Stoicism, which, among other things, advocated an acceptance of fate and an outlook on the world and events within it that, in common parlance, is often described as "philosophical." The other was Zeno of Elea (or Zeno the Eleatic). Elea was a town in southern Italy, and in the fifth and fourth centuries B.C., there lived here a group of philosophers, now known as the Eleatics, who taught that the cosmos consisted of a single, undifferentiated entity. In other words, there were, according to these thinkers, no separate objects or properties. The leading figure in this movement was Parmenides, who expressed his ideas in poetic form. Zeno, his follower and twenty-five years his junior (born 488 B.C.) is the more famous figure, and some of his paradoxes are widely known. Very little of Zeno's original writings survive, and almost all we know is derived from commentators such as Aristotle. Zeno did not, it seems, put forward any positive theories about the nature of the world, but instead constructed a series of ingenious paradoxes designed to demonstrate the absurdity of supposing that there was more than one thing in the universe, including "The Arrow" and (see following page) "Infinite Tasks."

Infinite tasks

Imagine again an object moving from one place to another. Suppose, for example, that you are walking from your home to the grocery store. In order to get to the store, however, you need to have reached and passed the mid-point, halfway between your starting point and finishing point. Let us suppose that the mid-point is the library. Now in order to reach the library, you need to have reached and passed the point midway between home and library—the end of your street. And to reach that point, you need to have reached and passed the point between your home and the end of the street. Clearly, this process could go on for ever. To travel any distance, however short, you need to have first traveled half that distance. But since the number of half-distances has no limit, that means that in order to walk from one place to another, you need to have covered an infinite number of half-distances. But no one can complete an infinite task (imagine counting to infinity).

As with the puzzle of the arrow, we could have used the example of any kind of change. Consider a kettle heating up. In order for the water to change in temperature from 20 degrees to boiling point, it needs to have reached a temperature of 60 degrees. But in order to reach that temperature, it must first have reached a temperature of 40 degrees, and so on. So in order to boil the water, a kettle must complete an infinite number of "tasks," in that it must go through an infinite number of temperatures. Since nothing can complete an infinite task, it seems to follow, again, that all change is impossible, and this threatens our belief in the reality of time.

When does the train leave?

A train is waiting at the station. At five o'clock precisely, the guard blows his whistle, and the train slowly moves out of the station. Now, as we would ordinarily think, there was a last moment at which the train was at rest, and a first moment at which the train is in motion (setting aside the difficulties we encountered a while ago when considering whether the arrow was moving at an instant). But the time between any two distinct moments can always be divided. In other words, between any two moments, there is always a third. But now suppose that there was indeed a last moment of rest. How could there be a first moment of motion? For whichever moment we alight upon as the first moment of motion, there is always one before it, and yet after the last moment of rest. But what could we say of a moment that was between the last moment of rest and the first moment of motion? That the train was *neither* at rest *nor* in motion then? It appears that either there was a last moment of rest, but no first moment of motion—in which case we cannot say that the train started to move at some point—or there was a first moment of motion, but no last moment of rest—in which case we cannot say that the train ceased to be stationary. Both of these are rather uncomfortable conclusions, and even if we settle on one of them, the choice just seems arbitrary. We ought to have a reason for preferring a last moment of rest over a first moment of motion, or vice versa.

Aristotle

Aristotle is one of the most important figures (some would say the most important) in the history of philosophy. His writings spanned an extraordinary range of topics—biology, astronomy, physics, music, art, and philosophy itself (although there is not in Aristotle's writings the very clear and self-conscious distinction that we would now recognize between "philosophy" and "science"). His thought in many areas, and especially physics and astronomy, remained dominant until the sixteenth century. This is all the more remarkable when one considers when he lived: he was born in 384 B.C., in Stagira in northern Greece. From 367 until 347, he was a member of an academy, which was something akin in terms of its intellectual aims to a university, although very much smaller. This was led by another towering figure in Greek philosophy, Plato. In 342, he was invited by King Philip of Macedonia to be tutor to his son, Alexander, later Alexander the Great. He returned to Athens for a while to found his own academy, and much of his writing from this period survives, apparently lecture notes, although they are much fuller and more polished than mere sketches. His discussion of the nature of time is a remarkably sophisticated one, and represents an enormous advance on earlier discussions. He articulated very clearly the notion of the infinite divisibility of time, and attempts to dissolve Zeno's paradoxes (including "The Arrow") by showing how they rest on questionable assumptions. One problem he introduces, and attempts to solve, is the issue about whether for every object that begins to move (or change, or to exist) there is a first moment of the new state and a last of the old, the problem discussed under "When does the train leave?" He also produced arguments against the idea that the world had a beginning.

Time atoms

The last two puzzles, although not the first, depend on the assumption that time is infinitely divisible: however short a period of time you are considering, you can always divide that period in half. There is, in other words, no shortest possible interval which still has some (non-zero) duration. But what happens if we abandon that assumption? What if we imagine periods of time to be composed of very, very small (but not durationless) intervals, which cannot be further divided? We might dub these "time atoms," after the original notion of an atom as an indivisible part of matter, as the word *atomos* is Greek for indivisible, but the name stuck even when it was discovered that atoms of matter are, after all, further divisible into subatomic particles.

The introduction of time atoms would certainly do away with the notion that any change in a finite period nevertheless involves an infinite number of steps. It would also allow us to say that there was a last moment at which the train was at rest, and a first moment at which it was in motion. But can we really make sense of time atoms?

Consider an object moving from one point in space to another. The object is moving so fast, and the distance between the points so small, that the journey takes exactly one atom of time; that is, it takes the shortest time possible. But, however fast the object was moving, it could always have moved more swiftly. Imagine, then, a swifter object doing the same journey. How long would its journey take? The answer must be less than one time-atom. But we have assumed this to be impossible! Would the existence of time atoms then imply that there was a limit to how fast things could move?

The disappearing present

How long is the present? We talk of the present century, but it would obviously be absurd to suppose that every year of this century is equally present. One is present, and the rest (as I write these words) future. But then the present year cannot be composed of equally present days. Only one day is present; earlier ones are past and later ones are future. And the day itself is further divisible into hours, some of which are past and some of which are future. The day is divisible into minutes, the minutes into seconds, and so on. So can the present take up any time at all? It seems not, since unless we believe in time-atoms, every period of time can be divided. But it would be absurd to divide the present into earlier and later parts, because what is earlier than the present is past, and what is later than the present is future. So the present takes no time at all; it is an instant, no sooner here than gone.

ené Magritte: *Time ansfixed* (1938); oil canvas.

But this conclusion is rather worrying, for two reasons. The first is that we commonly think of time as divided into past, present, and future. Indeed, there is nothing to time other than these three parts. But of the three, the past no longer exists, the future does not yet exist, and the present does not last long enough to exist! So time itself seems to have disappeared. The second reason has to do with the arrow puzzle. It is not an adequate answer to the arrow to dismiss the notion of an instant, a moment without duration, as an absurdity, or a mathematical fiction. For we all believe, surely, in the existence of the present moment; and the present moment, we have just seen, is an instant. And if we think of something changing, we think of it as changing in the present. If something is not capable of changing right now, then it is not capable of changing at all. But if the present takes up no time, then there is no time in the present for anything to change. So nothing changes in the present; so nothing changes at all.

William Hogarth: *Batho* *Manner of Sinking,* in *Sublime Pictures* (1764 etching and engraving.

The reality of past and future

It was noted just now, as if it were a matter beyond dispute, that the past no longer exists. But what does this mean? It certainly cannot mean (or can it?) that the past is just as unreal as the events of some fiction. We have, after all, memories of what happened a short time ago, and historical records of more distant events. We believe that at least some of these remembered or recorded events really happened. Something, in other words, corresponds to these present traces. Does that imply that somewhere the past is still going on? If it is still going on, then it cannot be the past, but the present. But if this is not implied, what entitles us to say that the past is real in a way in which the future is not? It is not enough to say that the past was once real, because it is equally true to say that the future will be real.

Perhaps the difference between the past and future is that only the past leaves its traces on the present. There are fossil remains from which we can deduce the existence at one time of giant reptiles. But nothing tells us of the kinds of species that will dominate life on Earth (if indeed there is any) in three hundred thousand years' time. Now if the past is nothing more than traces on the present, then when these traces are obliterated, what is now true of the past will cease to be true. Is that a conclusion we are willing to accept?

The river of time

Time is often compared to a river—an "ever-rolling stream." This metaphor is meant to capture time's ceaseless flow, the flow from future to present, and from present to past. But if the passage of time is a kind of movement, then it seems we are entitled to ask, How fast does time flow? Yet this is a very odd question to ask. We measure how fast things change by reference to time. Speed, for example, is distance covered per unit of time. In measuring the speed of time, then, we would be comparing time with itself. Five minutes takes...five minutes. What else could it take? But if five minutes necessarily takes five minutes, then time can neither speed up nor slow down. Yet it is surely an essential part of our idea of a process going at a certain rate, that it makes sense to imagine that rate changing. So if we cannot intelligibly talk of time speeding up, can we talk of the rate of the flow of time at all? And if we cannot talk of the rate of the flow of time, what meaning does talk of the flow of time have?

However, if time itself does not flow (the appearance of flow simply being a result of the way we are situated in time), then we have to treat all times—past, present, and future—as equally real. Or, in other words, the difference between past, present, and future would disappear.

Perhaps time could be measured against time if it had two dimensions. We ordinarily suppose that time has only one dimension, as opposed to space's three. Movement of one unit in one temporal dimension could be accompanied by a movement of two units in the other dimension. But this still requires us to make sense of movement in one dimension. And what difference to our ordinary experience of the world would it make if time had fewer, or more dimensions that it actually has?

St. Augustine

Augustine, who lived between A.D. 354 and 430 in Thagaste, Numidia (now Algeria) had, one may suppose, an intriguingly ambiguous upbringing: his mother was a Christian, but his father a pagan. In early adulthood, he rejected Christianity and led what he was later to describe as a rather dissolute life, although he was certainly already intellectually active, having secured when still young the position of Professor of Rhetoric at the University of Carthage. At the age of thirty-two, however, he was converted to Christianity. He was ordained priest not long afterwards, and in 395 was appointed Bishop of Hippo. He wrote a number of works ranging over theology, philosophy, and politics, including the intensely personal *Confessions*, in which he describes his early life and conversion to Christianity. In the course of this work he embarks on a long, rather tortuous, but fascinating discussion of the nature of time. The puzzle which initially motivates the discussion is that of what God was doing before he created the universe (the problem is clearly related to the one discussed on the next page under "Before the beginning"). He is exercised by a number of difficulties, including the one concerning the length of the present (see "The disappearing present"). His treatment of how it is we are able to measure time when what we are apparently measuring, a past event, no longer exists, leads him to the remarkable conclusion that time is something essentially mental. One possible interpretation of this idea is that it is the flow of time which is purely a mental construction, and that all times, whether past, present, or future, are as God sees them, and so equally real (see "The river of time," and "The reality of past and future").

At the end of ti
itself, old Father Ti
is unemploy

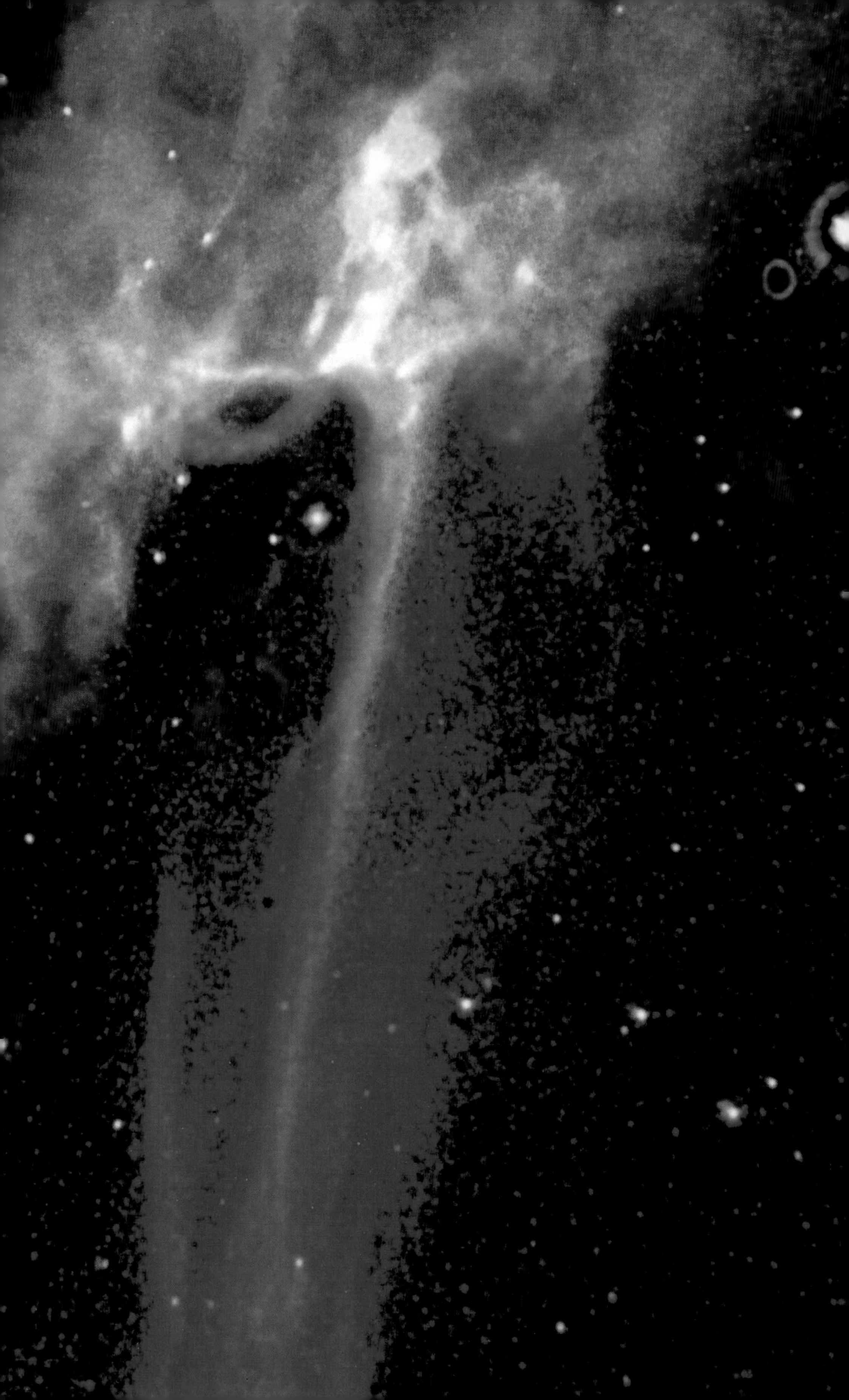

Before the beginning

According to some cosmologists, the universe began with the Big Bang, when an infinitesimally small amount of unimaginably dense matter suddenly exploded. Supposing this is correct, a natural reaction is to ask what was going on before the Big Bang. Perhaps nothing at all. But what does it mean to say that nothing happened before the Big Bang? It could mean that, for aeons of time, there was just empty (or virtually empty) space—a vast, featureless void in which nothing moved or changed in any way.

Thinking of the matter in this way requires us to imagine time existing when nothing whatsoever was happening. It means imagining time as a kind of container, independent of the things it contains. If we think of time in this way, then clearly we cannot say that time is the same thing as change, if "change" here refers to change in things (as opposed to the mere passage of time). The problem we now face is this: if there were indeed aeons of time before anything happened, we have no explanation of why the universe came into existence when it did. It is not simply that we cannot immediately put our hands on an explanation (although scientists may eventually come up with one)—there can be no explanation in this case. How could there be? When we want to know why a pipe burst when it did, we can refer to the previous cold snap. When we want to know why the dinosaurs died out, we can refer to the possibility of a previous collision between the earth and a meteorite. In other words, we look for changes in the environment that immediately preceded the event whose timing we are attempting to explain. But in the case of the Big Bang, we are supposing that nothing at all was going on before, so precluding any explanation. (Unless there is a kind of explanation that does not refer to prior causes.)

Those who think there should be an explanation for everything will not be happy with that outcome. However, there is another way of interpreting the phrase "nothing happened prior to the Big Bang," and that is that there was no time before the Big Bang. As one might put it, there was no before the Big Bang for anything to happen in. If that is what we mean, then we have an explanation of why the Big Bang happened when it did: The Big Bang represents year zero. The beginning of time is defined by the time of the Big Bang, so it makes no sense to ask why the universe did not come into being any earlier or later. Our difficulties are not yet over, however, because we have just introduced the idea of a beginning of time. Does this make sense? We talk of things beginning in time, but that implies that there was a time before they started. But there can be no time before the beginning of time.

he Cygnus Loop, one f the awe-inspiring ights of the universe. he light we see now as actually emitted ountless years ago.

The beginning and end of time

What is the present moment? It is the boundary between past and future. Without either the past or the future, the boundary could not exist (how could there be a boundary between something and nothing?). But now suppose, as we did just now, that there was a beginning to time, that there was a first moment. This boundary must at some stage have been present. But if it was present, then as the present is by definition the boundary between past and future, there must have been times which were already past when the first moment of time became present, and that is absurd. So we cannot consistently say both that the present is the boundary between past and future and that there was a first moment of time. There is a similar difficulty concerning the end of time: when it finally becomes present, there will still be a future. But then it cannot be the end of time after all! For there is no future at the end of time.

If we define the present as the boundary between past and future, then we cannot say that time will have a first or last moment, without any qualification. But perhaps this raises further difficulties.

The infinite past

If time had no beginning, then an infinite amount of time has already elapsed. If we think of the passage of time as the moving finger of the present alighting upon successive times in turn, then the present has moved through an infinite distance in order to reach the time at which you are reading these words. But, as we have noted before, it is impossible for anything to complete an infinite task (to do an infinite number of things). So if time really had no beginning, then the shifting present has done the impossible. Perhaps something can complete an infinite task, however, provided it has an infinite amount of time in which to do it. Suppose that, for all eternity, God has been counting down the number series. Yesterday, he counted down from one thousand to five hundred. Last year, he counted down from three hundred and sixty-five thousand to one hundred and eighty-two thousand, five hundred. The previous decade, he counted down from seven million, three hundred thousand to three million, six hundred and fifty thousand. And so on. The further back in time you go, the higher the numbers. Now, if there was no beginning to time, then there was no time at which God starting counting down—he has always been doing so. But today, he completed his task, counting down from five hundred to zero. So God has completed an infinite task. What then should we say? Because nothing by definition can complete an infinite task, time cannot have a beginning; or, unless time has no beginning, nothing can complete an infinite task?

Kant

Immanuel Kant (1724–1804) led, by all accounts, a life as austere as Augustine's must have been after his ordination. He was born in Königsberg in Prussia (now Kaliningrad, Russia), and went at the age of sixteen to the University there to study mathematics, physics, philosophy, and theology. In 1755, he began lecturing at the University, where he remained for the rest of his career, refusing offers of chairs elsewhere. Eventually, in 1770, he was appointed Professor of Logic and Metaphysics at Königsberg, finally retiring in 1797. He never married, and adhered to a daily routine of almost mechanical regularity, but was, according to contemporaries, both an inspiring lecturer and a warmhearted man. His *Critique of Pure Reason*, published in 1781, has been enormously influential. One section of the *Critique* is entitled "Antinomies of Pure Reason," and consists of arguments for contradictory conclusions set side by side. For example, he presents a proof that the world must have had a beginning in time, and a proof that the world must always have been in existence. The point of this rather surprising move is to show that the world as it is in itself (that is, independently of the way we perceive it) cannot be in time. Time, in other words, is something we project onto the world in order to make sense of our experience. The same also holds for space.

Changing the past and future

History can be rewritten, but can the events of which history speaks be altered? As things are, it would seem not. One expression of the direction of time is that we can only affect later events, never earlier ones. Suppose, however, we were able to travel back in time. Could we not then alter the course of history? Seeking to prevent the American Civil War, for instance, I travel back to the 1850s, to a time when I imagine there is the best chance of intervening in the course of events leading up to the conflict. Can I succeed? Here there is a paradox: If I do succeed, then it follows that there was no Civil War after all, and therefore there is no reason for me to have gone back in time to prevent it. The same problem can be more dramatically illustrated by the story of the time traveler who attempts to shoot dead his own grandparents before they reached maturity. If he succeeds, then he brings it about that he was never born (since his own conception depended on the earlier conception of his parents, which in turn depended on his grandparents' union). But if the time traveler was never born, he could hardly have succeeded in shooting his grandparents! Now if all time travel would necessarily result in the absurdity of changing the past, then it follows that time travel cannot be a possibility after all.

Similar reasoning about the future leads to a rather uncomfortable conclusion. Although we cannot (as yet) travel in the distant future, what was once future becomes present and so affectable by us. But we can no more change the future than we can change the past. Let us suppose that you have decided to go on a boat trip tomorrow. Since the decision is an entirely free one, you could change your mind at any moment. But if it is now already true that you will go on a boat trip, then you cannot now bring it about that you will not go on that boat trip. It is as much beyond your power to change what will happen as it is to change what has happened. Does that mean that, after all, we do not have freedom of choice? Or is it that, because we can still affect the future, there can be no truths about what will happen?

Going backwards

Could time suddenly start running backwards? Taken literally, this would involve all the events that have happened up to now taking place again, but in the reverse order, so what you did ten minutes ago would happen before the first moon landing. But that is absurd: a particular event can only happen once; if it happens again it is not that very same event, but rather one very like it. But if it is not the very same events that are happening again, then we have no reason to say that time is running backwards, only that the future will be a mirror-image of the past.

So perhaps we should rephrase our question. Could there be a world in which time went in the opposite direction from the direction it goes in in our world? To make sense of this, we have to think again about the kind of events that occur in time. For suppose that nothing existed except empty space. Then, if time had a direction at all, the past would still be earlier than the future (since the past is by definition earlier than the future), so there would be nothing to distinguish the direction of time in that empty world from the direction of time in this world. We may, however, imagine a world in which light travels from people's retinas to what we would ordinarily think of as light sources, like lamps, fires, the sun (although in the world we are imagining they would be light *sinks*: absorbing light rather than emitting it).

In this world, too, rivers would run uphill; people would become physiologically younger (but chronologically older); vacuum cleaners would throw out dust onto carpets; clouds of water vapor would contract and condense as they went down the spouts of kettles. We could imagine

me puzzles about e are as difficult to sp as the Mad tter's tea party.

witnessing such a world, but if we were part of it, would we see things in the order I have just described? If we were becoming physiologically younger, would that not mean that the traces of earlier events were obliterated? If so, then we would have no memories of what we had just seen, but would have memories of what we were about to see. At this point, the differences between the world we are imagining and the world we live in disappear. For in that world, the order in which things are perceived to occur is the same as the order in which they are perceived to occur in this world. So is there any room for the thought that, nevertheless, the real order of events is the opposite of the perceived order?

Glossary

archaeoastronomy The study of the practice and use of astronomy among past cultures. Orientations of prehistoric structures and alignments in the landscape are of primary interest, as well as other unwritten and written evidence.

astrolabe A medieval astronomical computer for solving problems relating to time and the position of the sun and stars in the sky, usually consisting of a disk and a pointer. By far the most popular type is the *planispheric astrolabe*, on which the celestial sphere is projected onto the plane of the equator. A typical astrolabe was made of brass and was about 6 inches (15 cm) in diameter.

azimuth Point on the horizon directly below the sun.

Big Bang The emergence of the universe from a state of extremely high temperature and density about ten billion years ago.

Bronze Age A phase in the development of material culture among the ancient peoples of Europe, Asia, and the Middle East marked by tools and weapons made of bronze replacing stone artifacts. In Egypt, Mesopotamia, Greece, and China the age began before 3000 B.C. It reached the British Isles by about 1900 B.C.

calendar round The basic structure of the pre-Columbian calendars of Meso-America. It consists of a ritual cycle cycle of 18,980 days, or fifty-two years of 365 days at the end of which a designated day recurs in the same position in the year.

Cambrian period Geological period lasting from 540 to 505 million years ago, often divided into the Early Cambrian epoch (540 to 520 million years ago), the Middle Cambrian epoch (520 to 512 million years ago), and the Late Cambrian epoch (512 to 505 million years ago).

catastrophism Explanation of the differences in fossil forms in successive stratigraphic levels as the product of repeated cataclysmic occurrences and repeated new creations.

chiliasm (from the Greek for 1000) A Christian belief held by some early Christians, by some seventeenth-century Christians, and by extreme Evangelicals today, usually taking the form that the earth will last 6,000 years before the Millennium is ushered in.

circadian (from the Latin for "about," *circa,* and "a day," *dies*) a term suggested by the biologist Franz Halberg in 1954 for "about a day."

circa-lunidian Approximating the periodicity of the lunar cycle.

circa-tidal Approximating the periodicity of the tides (about 12.4 hours).

coupled Synchronized in time by a direct interaction.

creationism Fundamentalist thesis that biological species owe their existence not to evolution but to special acts of creation by God.

Cretaceous period Geologic period from approximately 144 to 66.4 million years ago, with the first flowering plants, the extinction of dinosaurs, and extensive deposits of chalk.

cross-staff A medieval device for measuring angles of elevation. It consisted of a staff about three feet (one meter) long, fitted with a sliding crosspiece. Holding the staff to one eye, the operator would move the crosspiece until its lower end coincided with the horizon and its upper end with the point whose elevation was to be measured (for example, the top of a tower or a heavenly body). The altitude could then be read from a scale marked in degrees.

cursus Prehistoric earthwork consisting of two parallel ditches and banks.

dead reckoning Estimating a ship's position by keeping track of how far has been sailed and in which direction.

decans Thirty-six time-telling stars of ancient Egypt, whose morning risings were linked to the thirty-six administrative weeks of the year, and whose risings through the night marked the hours.

Deluge The Flood of Noah, which from 1650 to 1750 was thought to be the main geological cause of the strata. From 1780 to 1840 many geologists thought strata were deposited by a succession of deluges, Noah's being the last. This was also known as catastrophism or diluvialism. Lyell's uniformitarianism put paid to it. Recently, following Henry Morris, creationists argue that Noah's Deluge laid down all strata. Geologists reject this idea.

dispensationalism A form of chiliasm developed in the nineteenth century, including predictions of future events and a complex working out of the Second Coming of Christ.

entrainment Phase and period control of one rhythm by another.

evangelical A tradition stemming from the eighteenth-century Christian Revival putting especial emphasis on the Bible and the atoning death of Christ. Not all evangelicals are literalist in their use of the Bible.

equinox The time when the sun is overhead at noon on the equator. This occurs twice a year, roughly halfway between the solstices.

ethnocentrism The tendency to create a privileged view of our own culture in relation to others and to project our own way of comprehending things onto groups outside that culture.

free run Expression of a rhythm in an environment without time cues; a non-entrained rhythm.

genes Portions of DNA that code for unique proteins.

gnomon (from the Greek "to show or indicate") Any object that casts a shadow whose length or position is used to indicate time or geographical direction.

Gregorian calendar Calendar adopted under Pope Gregory XIII, now used everywhere in the world.

heliacal rising The first appearance of a star or constellation in the morning sky before dawn, after a period of being hidden by the sun.

heliacal setting The last appearance of a star or constellation in the evening sky after sunset, preceding a period of being hidden by the sun.

henge Prehistoric monument consisting of a circle of stone or wood uprights.

Hubble's Law Principle concerning the expansion of the universe expounded by the American Astronomer E.P. Hubble (1889–1953): the velocity of a galaxy is proportional to its distance from the earth.

Infinite Having no limit. The series of all even numbers, for example, is infinite in that, however large a given number, there is always a larger.

Instant Used variously to mean (a) a durationless point in time, (b) an infinitesimally small duration, (c) an indivisible but non-zero duration.

intertidal zone The portion of the coastline that is periodically flooded by the tides.

Iron Age Phase in the technological and cultural development of the Middle East, Asia, and Europe in which iron largely replaced bronze in implements and weapons. The age began in the Middle East and southeastern Europe about 1200 B.C., in China about 600 B.C.

intercalary month A month inserted in a year to harmonize a drifting calendar with timer markers, such as the seasons

Julian calendar Calendar introduced by Julius Caesar in 45 B.C.

literalism Taking Genesis One as teaching a six-day creation.

lunar month The 29.5 days between successive new moons.

lux A unit of measurement of illuminance, which is the amount of light falling upon a surface of defined area.

megalithic Made of or marked by the use of large stones.

melatonin A hormone produced by the pineal gland.

Mesolithic period The Middle Stone Age. Phase of human development characterized by chipped stone tools and microliths, very small stone tools intended for mounting together on a shaft. The age lasted from the last ice age to the start of agriculture.

Metonic cycle A nineteen-year cycle equalling a complete number (235) of lunar months, amounting to almost 6,940 days. The cycle is associated with the late fifth-century B.C. Greek astronomer Meton, but was known in Babylonia earlier in the same century.

Millennium The thousand years of Christ's reign after his Return, based on a literal reading of Revelation 20, which most Christians take as symbolic.

natural selection The process by which organisms evolve in response to challenges in their environments. Organisms that are well adapted to an environment have greater reproductive success, resulting in more offspring with similar adaptive characteristics.

Neolithic period The New Stone Age—phase of technological development characterized by stone tools shaped through polishing or grinding, by dependence on domesticated plants or animals, by settlement in permanent villages, and by the appearance of such crafts as pottery and weaving. The Neolithic followed the Palaeolithic period and preceded the Bronze Age.

neuron A nerve cell.

optic chiasm The point of entry of the optic nerves into the brain, where the nerves cross.

optic nerves The nerves conveying information from the retina.

Ordovician period Geologic period from 505 to 438 million years ago, with evidence of the first vertebrates and an abundance of marine invertebrates. The second period of the Palaeozoic era.

parapegma (Greek) A peghole star calendar.

Palaeolithic period The Old Stone Age—phase of human development, characterized by the use of rudimentary chipped stone tools.

Palaeozoic era Major interval of geologic from about 540 to about 245 million years ago, with an extraordinary diversification of marine animals ending in major extinctions.

periodicity The duration of a cycle.

Permian period Period of geologic time from 286 to 245 million years ago, with development of reptiles and amphibians, and deposits of sandstone.

phase Any point on a cycle.

Ptolemaic system Earth-centered model of the universe as perfected by the Alexandrine astronomer Ptolemy.

quadrant An instrument for measuring angles of altitude, comprising one-quarter of a brass disk and marked with a scale in 90 degrees.

quantum mechanics A branch of mechanics used in the study of elementary particles. It is based on Max Planck's theory of the emission and absorption of finite quanta of energy.

radiometric age-dating Technique for dating the age of mineral or rock by measuring the proportions within it of a radioactive element and the element into which it disintegrates at a fixed rate.

Recumbent stone circle Stone circle including a large stone placed on its side.

relativity theory The principle that all motion is relative. Einstein's Special Theory of Relativity (1905) adds to this the postulate that the velocity of light is always constant relative to an observer and derives important equations and predictions concerning gravitation and motion at constant velocity. His General Theory of Relativity (1916) deals in addition with acceleration.

reliquary receptacle, for especially holy relics.

Renaissance The period of rediscovery of ancient art and learning in Europe from the fourteenth to sixteenth centuries

retina Photoreceptors and neurons in the back of the eye that convert light energy into a patterned neural signal.

rhumb Any of the thirty-two traditional points of the compass.

rhythm Any process that repeats itself at approximately regular intervals.

sidereal month The time it takes the moon to move round the sky with respect to the stars. The mean sidereal month is 27·3 days.

solar day The time between successive sunrises or nightfalls, averaged over one year. Or, the sum of the duration of the day and night. This is determined by the speed of rotation of the earth about its axis (23 hours, 56 minutes), plus nearly 4 additional minutes attributed to the earth's rotation around the sun in the same direction. The solar day is actually 23 hours, 59 minutes and 56 seconds.

solstice The time when the sun reaches its furthest distance from the celestial equator. It is farthest north in June, when the sun is overhead at noon on the Tropic of Cancer, and farthest south in December, when the sun is overhead at noon on the Tropic of Capricorn.

spontaneous internal desynchronization The state of desynchrony that occurs when two or more rhythms within a single organism begin to express different periodicities.

suprachiasmatic nucleus A cluster of neurons in the hypothalamus, containing the circadian clock.

sun-compass orientation The use of the sun as a directional cue for navigation.

synodic month The phase cycle of the moon.

Time atom A non-zero period of time which cannot further be divided. The suggestion that time is actually made up of such atoms is a very controversial one, and is usually rejected.

tundra a vast level treeless Arctic region, often with a marshy surface and underlying permafrost. It covers about one-tenth of the total surface of the earth.

uniformitarianism In geology, the doctrine that existing processes acting in the same manner and with essentially the same intensity as at present are sufficient to account for all geologic change, i.e. that natural agents now at work on and within the Earth have operated with general uniformity through immensely long periods of time, and that explanations in terms of sudden catastrophes are not necessary.

Upper Palaeolithic period Around 30,000 B.C., *see* Palaeolithic.

Ussher chronology A chronology of the Old Testament, dating the creation of the universe at 4004 B.C., created by James Ussher (1581–1656), an Irish prelate of the Anglican Church.

volvelle A movable diagram made of rotating disks, designed to aid in the calculation of the locations of the planets throughout the year.

waning moon The moon during the decrease in the amount of its illuminated surface that is visible from the earth in the half month following a full moon.

waxing moon The moon during the increase in the amount of its illuminated surface that is visible from the earth in the half month following a new moon.

Zeitgeber (German for "time-giver") Any stimulus that can entrain a circadian rhythm.

index

Picture acknowledgments

Every effort has been made to trace all present copyright holders of the material used in this book, whether companies or individuals. Any omission is unintentional and we will be pleased to correct any errors in future editions of this book.

Cover by permission of the Houghton Library, Harvard University

p. 4 Private Collection/Bridgeman Art Library, p. 8 Torre dell'Orologio, Venice, Italy/Bridgeman Art Library, p. 11 Adam Woolfitt/CORBIS, p. 15 Mary Evans Picture Library, p. 16 Kevin Schafer/CORBIS, p. 18 Hammerby House, Uppsala, Sweden/Bridgeman Art Library, p. 20-21 Penny Brown

p. 22 Central Saint Martins College of Art and Design, London, UK/ Bridgeman Art Library

p. 25 Penny Brown, p. 28 Mary Evans Picture Library, p. 34 Penny Brown, p. 37 George McCarthy/CORBIS, p. 38 Roger Ressmeyer/CORBIS, p. 40 Bob Krist/CORBIS, p. 42 top Penny Brown, p. 42 bottom Penny Brown, p. 45 Michael Nicholson/CORBIS , p. 46 Penny Brown, p. 47 Penny Brown, p. 50 Penny Brown, p. 51 top Penny Brown, p. 51 bottom Penny Brown, p. 54 Archivo Iconografico, SA/CORBIS, p. 56 Private Collection/Stapleton Collection, UK/Bridgeman Art Library, p. 58-59 Penny Brown, p. 60 Mary Evans Picture Library, p. 64 Penny Brown, p. 65 Penny Brown, p. 66 Penny Brown, p. 67 Penny Brown, p. 68 Penny Brown, p. 69 Penny Brown, p. 70 Penny Brown, p. 71 top Penny Brown, p. 71 bottom Penny Brown, p. 72-73 Mary Evans Picture Library, p. 75 Penny Brown, p. 76 Penny Brown, p. 77 Penny Brown, p. 78 Penny Brown, p. 79 Penny Brown, p. 80 Mary Evans Picture Library, p. 83 Stapleton Collection, UK/Bridgeman Art Library, p. 84 Hulton-Deutsch Collection/CORBIS, p. 88 Mary Evans Picture Library, p. 93 Mary Evans Picture Library, p. 96-97 Archivo del Stato, Siena, Italy/Roger-Viollet/Bridgeman Art Library, p. 100 Mary Evans Picture Library/Explorer Archives, p. 103 Dennis di Cicco/CORBIS, p. 107 CORBIS, p. 108-109 Penny Brown, p. 111 National Library of Australia, Canberra, Australia/Bridgeman Art Library, p. 113 Gianni Dagli Orti/CORBIS, p. 115 Charles & Josette Lenars/CORBIS, p. 117 Penny Brown, p. 120 Mary Evans Picture Library, p. 123 Bettmann Archive/CORBIS, p. 125 Philippa Lewis/Edifice/CORBIS, p. 128 Adam Woolfitt/CORBIS, p. 130 Paul Almasy/CORBIS, p. 132 Duncombe Park, North Yorkshire, UK/Bridgeman Art Library, p. 137 Mary Evans Picture Library, p. 138 Michael Freeman/CORBIS, p. 139 Mary Evans Picture

Library, p. 140 Mary Evans Picture Library, p. 144 top Penny Brown, p. 144 bottom Roger Ressmeyer/CORBIS, p. 147 Mary Evans Picture Library, p. 149 Penny Brown, p. 152 by permission of the Houghton Library, Harvard University.

p. 154-155 Mary Evans Picture Library

p. 156 Duomo, Florence, Italy/Bridgeman Art Library

p. 159 School of African and Oriental Studies, London, UK/Bridgeman Art Library

p. 160 Mary Evans Picture Library, p. 163 The Worshipful Company of Clockmakers' Collection, UK/Bridgeman Art Library

p. 169 CORBIS , p. 171 Mary Evans Picture Library, p. 172 Michael Roberts, p. 175 Digital Art/CORBIS, p. 177 San Marco, Venice, Italy/Bridgeman Art Library, p. 180 Bettmann Archive/CORBIS, p. 183 Mary Evans Picture Library, p. 188 Mary Evans Picture Library, p. 192 Michael Roberts, p. 194 Punch Almanack 1882, p. 198 CORBIS , p. 200 Royal Society, London, UK/Bridgeman Art Library, p. 204 Eaton Gallery, Princes Arcade, London, UK/Bridgeman Art Library, p. 208 Penny Brown, p. 211 Penny Brown, p. 212 Penny Brown, p. 214 Mary Evans Picture Library, p. 216 Hulton-Deutsch Collection/CORBIS, p. 217 Penny Brown, p. 219 Penny Brown, p. 220 Gavin Graham Gallery, London, UK/Bridgeman Art Libray, p. 224 Penny Brown, p. 226 Stapleton Collection, UK/Bridgeman Art Library, p. 232 Art Institute of Chicago, IL, USA/Bridgeman Art Library. © ADAGP, Paris and DACS, London 2001. p. 234 The Trustees of the Weston Park Foundation, UK/Bridgeman Art Library, p. 237 Mary Evans Picture Library, p. 238 NASA/Roger Ressmeyer/CORBIS, p. 243 Private Collection/Bridgeman Art Library

Design concept: Broadbase
Design: Vladek Sczechter
Picture Research: Suzie Green

Sourcebooks, Inc.
P.O. Box 4410, Naperville, Illinois 60567–4410

Tel: (630) 961–3900
Fax: (630) 961–2168

Library of Congress Cataloging-in-Publication Data

The discovery of time / edited by Stuart McCready.
p. cm.
Includes bibliographical references and index.
ISBN 1-57071-675-7 (alk. paper)
1. Time. 2. Time measurements. I. McCready, Stuart.

QB209 .D57 2001
529'.7--dc21 00-067037

Printed and bound in Italy

1 2 3 4 5 6 7 8 9 0

ISBN: 1-57071-675-7